COSMIC RAYS

THE WYKEHAM SCIENCE SERIES

General Editors:

PROFESSOR SIR NEVILL MOTT, F.R.S.
Emeritus Cavendish Professor of Physics
University of Cambridge

G. R. NOAKES
Formerly Senior Physics Master
Uppingham School

The aim of the Wykeham Science Series is to introduce the present state of the many fields of study within science to students approaching or starting their careers in University, Polytechnic, or College of Technology. Each book seeks to reinforce the link between school and higher education, and the main author, a distinguished worker or teacher in the field, is assisted by an experienced sixth form schoolmaster.

COSMIC RAYS

J. G. Wilson
The University of Leeds

WYKEHAM PUBLICATIONS (LONDON) LTD
LONDON AND WINCHESTER
SPRINGER-VERLAG NEW YORK INC.
1976

Sole Distributors for the Western Hemisphere
SPRINGER-VERLAG NEW YORK INC. NEW YORK

First published 1976 by Wykeham Publications (London) Ltd.

Cover illustration—Smashing of a silver atom by a 30,000–*Gev cosmic
ray.   The collision of the cosmic particle with the silver nucleus
produced ninety-five nuclear fragments whose tracks form the star.*

*Printed in Great Britain by Taylor & Francis (Printers) Ltd.
Rankine Road, Basingstoke, Hants.*

Library of Congress Cataloging in Publication Data

Wilson, John Graham, 1911–
    Cosmic rays.
    (The Wykeham science series; 40)
    Includes index
    1.   Cosmic Rays.
QC485.W483   539.7'223   75-38743
ISBN 0-387-91131-6

PREFACE

THE history of our understanding of the cosmic particle radiation is notable because a great deal depends on the development and interrelation of straightforward concepts rather than on abstruse specialities in particular areas of physics. The theme of this work is centred on this aspect of ' cosmic rays ' and so it is not specifically directed at the committed physicist: I hope it will be of interest to many scientifically-inclined sixth-form students and undergraduates, perhaps even to people whose first commitment is in quite different disciplines.

I am grateful to the Carnegie Institution of Washington, the Physics Department of the University of Durham, Professor P. H. Fowler, Dr. C. J. Hatton and Dr. A. L. Hodson for permission to reproduce photographs, and particularly to Dr. Hatton for his clear judgement in giving advice on certain chapters.

Above all I am grateful to my collaborator, Mr. G. E. Perry, for his patience, care and invaluable guidance in the preparation of the manuscript and specially for his encouragement in developing the general form of the book.

Leeds   J. G. WILSON
March 1975

UNITS AND SYMBOLS

THE internationally recognized standard system of units (the S.I. system) is derived from seven base units, independently defined. In this book we use only those of length, mass, time, electric current and temperature. The metre and the second are defined in terms of a specified wavelength *in vacuo* and the frequency of a suitably repro-ducible atomic radiation, the kilogram is defined as the mass of a platinum–iridium cylinder kept at Sèvres in France, and the ampere is defined in terms of the force between infinitely thin and long parallel conductors each of which carries the specified current. The kelvin is defined relative to the absolute zero of temperature by a reproducible state, the triple-point of water. All of these are defined in a manner which makes the present accepted magnitude close to that well-established in earlier systems. From them other, derived, units have been identified and named.

S.I. Units

Physical Quantity	Unit	Symbol
length	metre	m
mass	kilogram	kg
time	second	s
electric current	ampere	A
temperature	kelvin	K
force	newton	$N = kg\ ms^{-2}$
energy	joule	$J = kg\ m^2\ s^{-2}$
power	watt	$W = kg\ m^2\ s^{-3}$
electric charge	coulomb	$C = As$
electric potential difference	volt	$V = J\ A^{-1}\ s^{-1}$
magnetic flux density	tesla	$T = V\ s\ m^{-2}$
frequency	hertz	$Hz = s^{-1}$
solid angle	steradian	sr

Since quantities which have to be expressed in this system may often be many orders of magnitude larger or smaller than the unit, prefixes characterizing fractions or multiples have been adopted. The most important of these represents steps of 10^3 in magnitude: those used in this book are

10^{-9} (of the unit)	nano	symbol	n
10^{-6}	micro		μ
10^{-3}	milli		m
10^3	kilo		k
10^6	mega		M
10^9	giga		G
10^{12}	tera		T

One or two other multiples have long histories and will certainly continue in use. The most important are the centimetre (cm) $= 10^{-2}$ m, the ångström (Å) $= 10^{-10}$ m. Others, which may cease to be used more rapidly are the gauss, $G = 10^{-4}$ T and 'γ' $= 10^{-9}$ T.

Other Units

For the foreseeable future it will be necessary and important to understand work which uses other systems of units, and while the dependence upon these has been reduced to a minimum in this book, some must be noted.

The astronomical unit—the mean distance from the earth to the sun is about $1{\cdot}5 \times 10^{11}$ m.

(The light year—the distance travelled by light *in vacuo* in one astronomical year is about 10^{16} m.)

The parsec (parallax-second)—the distance at which the mean radius of the Earth's orbit, viewed normally, subtends an angle of one arcsecond—about 3×10^{16} m.

Angular Coordinates

A variety of uses are made in cosmic ray physics of the angular element of polar coordinates: the radial distance in the complete coordinate specification is very rarely significant and, indeed, is often meaningless.

(i) The *local* coordinates—angle from zenith, θ; the angle in azimuth ϕ, normally measured from geographic north.

(ii) Terrestrial coordinates—(latitude and longitude).

(iii) '*Celestial*' coordinates are defined relative to the rotation axis of the earth by an azimuthal angle—the Right Ascension (R.A.), λ, and a latitude angle—the declination, δ. Declination is measured in degrees (conventionally from $+90°$ in the north polar direction to $-90°$ in the south polar direction); Right Ascension is *either* measured in angular degrees *or* in time, since it is experimentally determined as a time between local sidereal time and the observed time of transit of the local meridian by the point described (in classical astronomy, a star).

(iv) *Galactic* coordinates are described relative to the plane of circular symmetry of the galaxy: they define, in particular, directions to the galactic poles. The N galactic pole is now defined from observations of 21 cm radio-emission in the galactic plane: it lies at R.A. 12 h 49 m, $\delta + 27\cdot4°$.

Time

The strict definition of the second has been indicated above. In fact it differs only slightly from that defined from a solar day—the time of the rotation of the earth about its axis relative to the earth–sun line.

Local (solar time) is based on local noon and so varies around the earth. A common time, in which events in different parts of the earth can be synchronized, is based on Greenwich Mean Time; this is frequently now referred to as Universal Time (UT) and this form is used where necessary in this book.

Sidereal time is based on the rotation-period of the earth relative to distant stars and differs from solar time by approximately 1/365. Right Ascension is measured in sidereal time.

CONTENTS

The Cosmic Ray Observatory, on the slopes of Mt. Chacaltaya in the Bolivian Andes at an altitude of 5220 (17 100 ft), provides the highest permanent facility for cosmic ray experiments in the world. It is used by workers from many other countries as well as by Bolivians.

Work began in the 1950's, and the observatory is now comprehensively equipped for detailed studies of Extensive Air Showers at an earlier stage of development than is possible at sea-level. The facilities also include a neutron monitor, directional muon telescopes and underground muon detectors for studies of time variations of cosmic ray intensity. In this work Chacaltaya has a particularly important position, since it is at an altitude where counting rates are high, and statistical precision accordingly good, and also close to the geomagnetic equator, with a consequent absence of low momentum ($\sim$13 GV for vertical incidence) primary particles.

CHAPTER 1

discovery and identification of a ' cosmic radiation '

1.1 *Introduction*

THE flow of charged particles which continually bombards the earth from outside is one of the new sources of knowledge which have changed our ways of thought on astrophysical problems. These particles, each with very high kinetic energy, are what we call ' cosmic rays '. To optical astronomy, which has monopolized exploration outside the earth for hundreds of years, we are now able to add radio-astronomy, X-ray astronomy and ' charged-particle astronomy ' or ' cosmic rays '. Each has provided its own surprises, but these different features are complementary and we are now reaching a stage when they fit together into a convincing overall structure.

This book is mainly about ' cosmic rays ' as an ingredient of the new astronomy, and so a great deal of it has got to be about quite recent discoveries, and the way in which these link up. However this very modern material will be prefaced by an historical section, for the subject of cosmic rays is one of the best examples there is of the development of physical ideas from simple, almost casual beginnings to the point where they fit together with others which have themselves come from quite different sources. It illustrates vividly how physics really grows.

It was the distinguished French physicist Charles Augustin Coulomb who, near the end of the eighteenth century, first seems to have noticed the property which later led people to the discovery of cosmic rays. He observed that a charged sphere hung by a long silk thread gradually lost its charge, and he suspected that it was more likely that the air is slightly conducting than that this leakage took place along the silk. But this was an observation in isolation; he and his contemporaries did not have any picture of the way in which electricity is conducted which would allow it to be ' explained '.

At the beginning of the present century the situation was quite different. Ions had been recognized, and agents such as X-rays and the radiations from radioactive substances were known to make air strongly conducting. Measuring instruments, too, were becoming well developed which could record how quickly charge leaked away from the isolated electrode of what came to be described as an ' ion chamber ' and which was the quantitative analogue of Coulomb's charged sphere. A simple ion chamber containing its own measuring

1

device was the gold leaf electroscope, and it was in such an electroscope that a slight change of design made conduction through the air certainly distinguishable from conduction over the insulating holder of the central electrode and its leaves. The other side of this insulator was maintained at the original potential which had been given to the leaves (fig. 1): any leak across the insulation would thus prevent loss of charge from the leaves, not contribute to it. A simple modification, yet decisive in the confidence which it brought to observations.

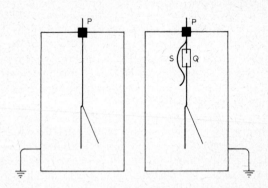

Fig. 1. The gold-leaf electroscope. Left: the original form of the instrument in which the rod carrying the gold leaf (which is repelled when the rod is charged) is insulated from the earthed case by the insulator P. Here leakage of charge through ionized gas in the chamber cannot be distinguished from leakage over the surface of the insulator P. Right: the improved instrument from which this uncertainty has been eliminated. The rod through the insulator P is maintained at the charging potential and is separated from the section of rod carrying the gold leaf by a second insulator Q. This sector and the leaf are charged by momentary contact of the spring S using an external magnet. Any leakage across Q will be in a direction to maintain the charge on the leaf and not to discharge it. Discharge must take place through the gas of the electroscope.

When radioactive preparations were brought near an electroscope or when X-rays fell upon it, leakage of charge took place in the electroscope and it was understood that what was happening was 'ionization' of the air in the instrument. From neutral molecules, positive and negative 'ions' were being produced, as a rule by an electron being removed from a neutral molecule, leaving a positive ion, and this electron attaching itself to another molecule forming a negative ion. These ions then moved in the electric field towards the electrode or towards the case of the electroscope according to their charge, and, one electronic charge at a time, the electroscope was discharged. When known sources were kept away from an electroscope, the loss of charge almost, but not quite, stopped. It is an indication that the problem was already seen in quantitative terms that at this stage it was established that the residual leakage of charge corresponded to the release of something like ten pairs of ions per second in each cubic centimetre of gas in the ion chamber or electroscope.

A natural explanation of this effect was to attribute it to weak radio-active radiations from ordinary matter around the electroscope. Radio-active elements are not abundant anywhere but they are very widespread indeed at extremely low concentrations. What was rather surprising was that it did not seem possible to alter the rate of leakage much by moving the instrument about. Serious observations of this pheno-menon began in the year 1900 when two German physicists, J. Elster and H. Geitel, and in this country C. T. R. Wilson, best known as the inventor of the cloud chamber, quite independently came to similar conclusions. The leakage was the same in daylight and in the dark, whether the charge on the measuring electrode was positive or negative, and whether the initial potential was high or low. It was Wilson who showed that at normal pressure about ten ions of each sign were produced per second in each cubic centimetre and that this rate was proportional to the pressure of air. He then went on in an historic sentence: ' Experiments were now carried out to test whether the production of ions in dust-free air could be explained as being due to radiation from sources outside our atmosphere, probably radiation like Röntgen rays or cathode rays, but of enormously greater penetrating power! ' He took his electroscope to a remote railway tunnel in Scotland, where trains did not run at night, and there in the centre of the tunnel made further leakage measurements. There was no great difference between the measurements here and at the surface, and he concluded therefore that the hypothesis of an extra-terrestrial source was unlikely.

1.2. *The exploration of the atmosphere*

But the situation remained puzzling. When an electroscope was taken by boat on a snow-fed mountain lake, over water which must surely have been much less radioactive than most matter, there was, it is true, some diminution of the effect, but not very much. It was not until more than ten years later that the ultimate effort was made to take the electroscope right away from the immediate surroundings of terrestrial rock and water, and first V. F. Hess and then other observers embarked upon projects to take ion chambers high into the air on balloons. The results were revealing and convincing. At first the leakage became smaller as the balloon went up, but at a height of perhaps a kilometre the effect began to change, and beyond that the leakage of the electroscope increased continuously with

3

increasing altitude.   Hess himself carried instruments to a height of about 5 km, and later W. Kolhörster went further, to about 9 km where the rate of leakage was about ten times that at the ground.

Kolhörster's flights were carried out very shortly before the outbreak of the first World War; and it was to be nearly twenty years before work extended knowledge to significantly greater heights.   In the early 1930s A. Piccard and M. Cosyns reached an altitude of 16 km in a manned balloon, but to do this they had to make the flight in a sealed, pressurized gondola, and several workers at more or less the same time came to realize that the human factor was becoming the limiting constraint, and attention turned to automatic recording systems which might be sent much higher and then bring back data to the ground.   Of several workers, Victor Regener was outstanding; he flew an automatic recording ion-chamber to an altitude of 25 km. The early part of his curve agreed closely with that of Kolhörster, but towards his limiting height the rate of increase of intensity became slow, and indeed the intensity seemed to be flattening out at a steady value.   Could it be that this flight had reached the source of cosmic radiation?

The culmination of this upward exploration using balloons came in 1936, when G. Pfotzer flew apparatus from Stuttgart to an altitude of about 30 km.   He showed that the region reached by Regener was not a plateau, but somewhere very near to a maximum value, and that at still greater heights the measured intensity was found to decrease.   Pfotzer's detecting apparatus was, in fact, not an ion-chamber, as had been used by his predecessors, but rather a Geiger-counter telescope.   This detector was not omnidirectional in its sensitivity, as an ion-chamber is, but only sensitive to charged particles moving in directions not greater than about 20° from the zenith, and so what is measured was actually something rather different.   However the effect is qualitatively the same, and so the three critical stages in determining the altitude-variation of cosmic ray intensity are shown together in fig. 2.

Up to this point the development of cosmic ray physics had not drawn very much on what was going on in other branches of physics, and in particular on such things as the detailed properties of the ionizing radiations encountered in studies of radioactivity.   But once it became clear that some ionizing agency seemed to be affecting the atmosphere from beyond it rather than from the earth below, its relationship to the known radioactive radiations became a matter of obvious interest.   It was perhaps inevitable that 'cosmic rays' should at first be thought of as like γ-radiation, for this was the most penetrating form of radioactive radiation, and the new, extra-terrestrial radiation seemed even more penetrating.   Partly because the absorption of γ-rays (i.e. of high energy electromagnetic radiation)

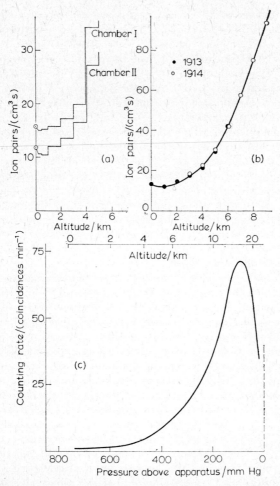

Fig. 2. Altitude variation of ionization. (a) Balloon ascent by Hess (1912) carrying two ion chambers. (b) Ascents by Kolhörster (1913, 1914) using ion chambers. (c) Coincidence counter telescope flown by Pfotzer (1936).

was incompletely understood, with ' pair production ' which (see p. 28) we now regard as an important mode of absorption of γ-rays completely unknown, and partly also because this assumption was built upon energetically by the formidable R. A. Millikan, the theory of ' ultra-γ-rays ' held the field until about 1930. Millikan undertook painstaking studies of the absorption of the new radiation, particularly in the waters of a high altitude snow-fed lake, and tried to reproduce his observations as the superposition of the absorption of a few identifiable ' ultra-γ-rays ' (see the phrase (p. 8) used by Skobelzyn), and speculated on their origin. He tentatively identified them with the energy of formation, from protons, of more complex nuclei, and his series extended up to a ' γ-ray ' energy of about 200 MeV*. Extreme assumptions were obviously being made, involving energies far beyond those encountered from terrestrial materials or justified by calibrations in the laboratory. No-one dreamt of the very much larger actual energies which charged particle primaries indeed carry. Millikan's bold approach to the absorption of cosmic rays was finally superseded rather suddenly. The pair production process, which limits the penetration of γ-rays with increasing energy, became qualitatively recognized several years before any theory of it was possible, and also at about the same time experiments began to be carried out that demonstrated that ionizing particles travelling long distances were an important feature of the cosmic rays.

1.3. Counter telescopes

This became possible when the German physicists Bothe and Kolhörster used Geiger counters to detect the ionization due to cosmic radiation and, in particular, worked out ways of using the simultaneous discharge of two or more separate counters to identify the passage of a single charged particle through a number of counters. This technique is the basis of counting methods that have played an important part right through the history of cosmic ray studies up to the present day. It is known as ' coincidence counting ', and a set Geiger counters placed so that it is possible for a single particle to pass through all of them in a straight line is described as a ' counter telescope ' (fig. 3). The coincidences of discharge between three counters arranged as a ' telescope ' (fig. 3 (a)) are reduced to almost nothing if the central counter is displaced out of the beam of the telescope (fig. 3 (b)), and the interpretation of telescope operation as arising from the passage of a charged particle following a straight path was fully established when Blackett and Occhialini in 1932 used the coincident discharge of a counter telescope to operate a cloud chamber placed to intercept the beam defined by the telescope, and succeeded in photographing the trajectories of particles (fig. 3 (c)).

With counter telescopes it was not only possible to show that what we were observing at the ground were in fact charged particles, but

---

* The electron volt (eV), the energy change when an electron charge is taken through a difference of potential of one volt, is approximately $1\cdot6 \times 10^{-19}$ J. Particle energies discussed in this book will normally be many orders of magnitude greater than one electron volt, thus 1 MeV ($= 10^6$ eV), 1 GeV ($= 10^9$ eV) will often be used, and at very great energies the use of a simple index (e.g. 10^{20} eV) is common practice.

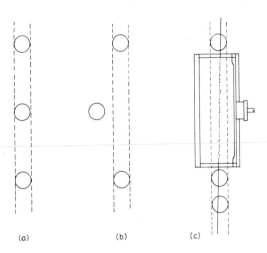

(a)   (b)   (c)

Fig. 3. Counter telescopes. In (a) three Geiger counters (seen end-on) are
placed so that a single particle can pass through all three and discharge
them in coincidence. If one counter (b) is moved out of the line of the
other two, the three counters can no longer be discharged simultaneously
by a single particle, and the arrangement is no longer a telescope
identifying individual particles in a particular direction. The intro-
duction (c) of a cloud chamber to enclose a section of the beam defined
by a counter telescope and operated by the discharge of the three counters
in coincidence, confirmed the hypothesis of telescope operation, almost
always showing a single track photograph in the chamber. It also
provided a technique, universally adopted, for photographing tracks of
cosmic ray particles both in simple and complex situations (see fig. 29).

also to observe the direction in which these particles came. It was
very soon shown, using such a telescope, that these particles came
mainly from vertically above, and that the intensity fell to almost
nothing as the direction of the telescope was changed to point
horizontally. Thus there was confirmation, if any were needed, that
particles were coming down from the high atmosphere and that as
one tilted the telescope to look through increasingly thick layers of
air the number of particles quickly became small (fig. 4).

1.4. *Showers*

However, experiments with a telescope tilted near to the horizontal
led to further, quite unexpected, observations which were to take the
subject forward through another critical step. It was found that a
set of Geiger counters operated in coincidence but placed so that a
single particle moving in a straight line could not possibly go through
them all was nevertheless from time-to-time discharged simultaneously.
One could only conclude that the particles which account for cosmic

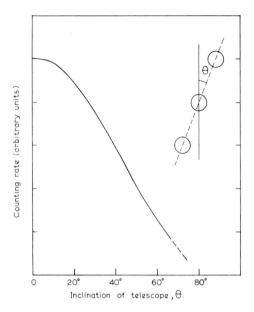

Fig. 4. Variation of counting rate of a telescope as it is tilted from the zenith $(\theta = 0°)$ towards the horizontal.

ray phenomena at the surface of the earth do not always arrive alone but at least sometimes as members of some sort of group, and visualizing these particles coming more or less vertically, it is not surprising that these groups of particles coming down together almost at once became known as 'cosmic ray showers'.

The very first observations of showers were in fact not made using counter systems and did not excite much attention. In 1929 D. Skobelzyn reported work which he had carried out at Leningrad on cloud chamber tracks of β-particles photographed in a fairly strong magnetic field (1500 gauss = 0·15 T). A few of his photographs showed tracks of particles moving in near-vertical directions and thus right out of the plane to which his source-particles were constrained. These were not counter controlled exposures and although one photograph showed three of these anomalous tracks they could not be considered as established to be simultaneous. However Skobelzyn speculated whether these might not be particles secondary to the 'Hessschen Ultra-γ-Strahlen'. He was almost certainly correct as to the source of his near-vertical tracks if not as to the full nature of their source.

It was not until 1938 that the full importance of 'showers' began to be evident. In that year the very distinguished French physicist

8

Pierre Auger, with two young collaborators, measured the rate of coincidence between a fixed (vertical) counter telescope and a third counter, as this was moved from within 1 m of the telescope to much greater distances. The rate of coincidences fell off as this distance was increased, but it was still measurable at 75 m separation. This scale of the lateral spread of showers (which in later experiments was to prove to be very much greater even than this) earned them the name by which they are often still known—' extensive showers ' or ' extensive air showers ', with the widely used abbreviation ' EAS '.

Another feature was shortly afterwards revealed in counter-controlled cloud chamber photographs. In 1939 A. C. B. Lovell and J. G. Wilson operated two cloud chambers rather more than 5 m apart so that they were triggered by a ' shower-sensitive ' system of counters in coincidence. Each chamber offered an area of about 0·01 m² and in a number of photographs each showed the tracks of many particles. So these ' showers ' could also be intense events with large numbers of particles arriving all at the same time, and, by implication, involving a very large aggregate amount of energy. Lovell and Wilson put the energy of the showers in these photographs, taking the most conservative estimate, as greater than 1 TeV (10^{12} eV), and thought that the total energy actually involved could probably be as much as one hundred times greater. At that time, this was a really startlingly large amount of energy.

One particularly important shower experiment had been carried out in 1933 by the distinguished Italian physicist Bruno Rossi. He took a small group of Geiger counters, not in a straight line, and proceeded to measure the absorption of the groups of particles which appeared to discharge these counters simultaneously. He measured the rate at which this system counted ' showers ' as he placed increasing thicknesses of lead, which he had chosen as a promising absorber, above the counter arrangement. He found that what happened was not a simple absorption at all: in fact the first one or two centimetres of lead, far from cutting the rate of showers counted by the Geiger counters below, increased it quite noticeably. This was such an extraordinary result, and so contrary to ' common sense ', that at least one distinguished journal is said to have declined to publish anything so eccentric. However, the reality of this effect, fig. 5, was quickly established, and it became clear that thin layers of lead were not so much acting as absorbers of existing showers, but, more effectively, as sources of new ones.

This experiment was in fact a quite crucial one. If it were possible to make new showers, originating in layers of lead in the laboratory, was it not likely that all showers were something that had an origin within the atmosphere, and was it not likely, too, that many other particles, observed singly in a Geiger counter telescope had the

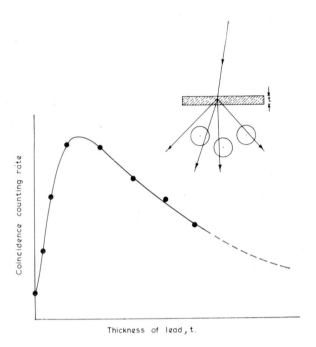

Fig. 5. The Rossi experiment, in which the coincidence counting rate of a
group of counters so placed that several particles are required to
discharge them is studied as a function of thickness, t, of a layer of
lead placed above them. The maximum rate is found at a thickness of
about 20 mm. The inset indicates the proposed mechanism of locally-
produced showers.

same kind of origin, even although they were observed as individuals
rather than members of a shower group?

1.5. *Primary and secondary radiation*

So the picture became established that while cosmic radiation had
its origin outside the earth, some at least of the particles observed in
counter systems were of more local origin, and, in all probability,
secondary to the extraterrestrial radiation. We are introduced to the
notion of a ' primary radiation ' which comes to the earth from
outside, and a ' secondary radiation ' which is the product of primary
radiation as it comes into the matter of the atmosphere, or the absorbing layers of Rossi's experiments. From this point of view it is
interesting to look back at fig. 2 (c) which shows the result of high
unmanned balloon flights and which shows that the intensity of cosmic
radiation in the atmosphere reaches a maximum near the top and then

at still greater heights falls away again. Taken in combination these suggest strongly that a great deal, and perhaps all, of what we observe in the atmosphere, and certainly on the ground, is only a secondary radiation and that up to this time contact with the primary radiation had been only at second-hand.

1.6. The primary radiation

The basic discoveries about the primary radiation were of a general kind, stemming from existing experience about radioactive radiations. Were the primaries charged particles, or were they (uncharged) radiation, presumably electromagnetic? This was the question which had been asked and answered about α-, β- and γ-rays a quarter of a century earlier, and now it came up again, but on a different scale and in quite a different environment.

The so-called ' Lorentz force ' deflects the trajectory of a charged particle moving in a region of magnetic field, and in elementary physics one usually only comes across the two extreme cases of motion in a uniform field: motion along the direction of the field, when the force is zero and the path of the charged particle undeflected, and motion perpendicular to the field when the force, uniform in magnitude and perpendicular to the instantaneous direction of motion, results in a circular trajectory. Other modes of motion in a uniform field may be thought of as derived by taking two velocity components respectively along and perpendicular to the direction of the field; these all result in a spiral motion of constant pitch angle.

The motion which would be expected of charged particles (if that is what we are dealing with) in the earth's magnetic field is much more complicated. Even if we regard this field as strictly that of a magnetic dipole (and it is only approximately of this simple form) such a field is uniform neither in direction nor magnitude, and charged particles in it can follow paths of the most extraordinary complexity. Very few of these can be represented by straightforward analytical expressions, and almost always they have had to be determined by numerical methods which, until the development of modern computers, were very tedious indeed. In spite of this, however, when the question arose of using such calculations to test whether cosmic-ray primaries were or were not charged particles, studies of the problem were not completely novel: a Norwegian physicist, C. Störmer, had devoted many years to the examination of such trajectories as part of his pioneering work on a theory of aurorae.

But for the critical question about the nature of primary particles, we can understand the main conclusions by considering two extreme and quite simple situations: the motion of a charged particle approaching strictly along the axis of the dipole, and that of a particle moving strictly in the equatorial plane of that dipole.

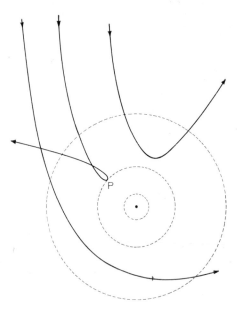

Fig. 6.   Trajectories of particles of the same charge and momentum moving in
the equatorial plane of the geomagnetic dipole field.   For a discussion
of the significance of the broken circles, see the text.

As to the first of these, it is evident that whatever the nature, the
charge or the velocity of the particle, it will undergo no deflecting
force and will continue along its straight trajectory until it reaches the
earth.   For the particle moving in the equatorial plane, the motion of
a charged particle is not force-free, but at least the force acting on the
particle is also always in the equatorial plane and so the particle
remains in that plane.   The general form of the motion is illustrated
in fig. 6 which shows three typical trajectories for similar particles
(that is to say, particles of the same nature, charge and velocity) the
paths of which were initially parallel, but which were ' aimed '
differently relative to the position of the dipole.   There are a number
of important features which we can understand from this diagram:
(i) the curvature of each trajectory becomes stronger as it approaches
the dipole, for the field increases (in fact, proportionally to r^{-3}) as one
gets near to it.   Since the curvature would ultimately become infinite,
the trajectory never under any circumstances reaches the dipole.
(ii) The trajectory accordingly has a point of closest approach to the
dipole and it is symmetrical in shape about this point.   The diagram
is drawn for particles of a particular sign of charge.   Particles of
opposite charge would move along trajectories similar but oppositley

12

curved. The trajectory is tangential at the point of closest approach. (iii) The actual distance of the point of closest approach depends upon a number of factors. As is clear from the diagram it depends upon the 'aiming point' for each individual trajectory, but some other factors are not difficult to understand. It will certainly depend on the strength of the dipole, and upon the charge, the mass and the velocity of the particles we are considering. The last two features can be brought together as the momentum of these particles, and so for an actual problem, with the strength of the dipole fixed (for example, that of the earth) and considering particles all of the same charge, the distance of closest approach is determined by the momentum of the particle.*

Bearing in mind that we are taking the diagram to refer to the (fixed) dipole moment of the earth, we can now regard fig. 6 as of variable scale; just what real distance any distance in the diagram represents depends upon the momentum which we assume for the approaching particle. So we can represent the earth on the diagram by a circle of radius which we vary to suit the different assumed momenta of the approaching particles, and in fig. 6 three 'earth circles' have been drawn which correspond to three possible momentum values of the incoming particle. The largest circle corresponds to a rather high momentum and it indicates that particles with that momentum reach the earth at the equator quite easily. The smallest circle corresponds to a much lower momentum and here it appears likely that such particles cannot reach the equator at all. The intermediate circle corresponds to a momentum for which it is just possible, but difficult, for the particles to reach the earth in the equatorial plane.

So it follows that in the idealized model, assuming the earth to have a perfect dipole field, all charged particles of whatever momentum can reach the earth along the axis of the dipole (the 'geomagnetic axis' of the earth) while only those with more than a certain amount of momentum can reach it in the equatorial plane of the dipole (the 'geomagnetic equator'). The earth's actual field is not very different from that of a perfect dipole, and we would expect this result to be substantially true for the real situation. We also expect (and this can be established by computation) that there will be a gradual transition from one extreme to the other, and that the limiting momentum for a particle to be able to reach the earth will increase as we go from the pole to the equator.

The bearing of this discussion on the nature of primary cosmic radiation will now be apparent: if the primary radiation consists of

* See Appendix A, which discusses the basic features of magnetic deflection in more detail.

charged particles, then unless they are *all* of very high momentum, *more* should reach the poles (and high latitudes in general) than reach equatorial latitudes. Clearly there should be no such effect if the primaries are uncharged particles or electromagnetic radiation. Do measurements of cosmic ray intensity show such an effect?

To test this critical, if simple, question involves experiment on a new dimension. It is something to be done on a world scale with

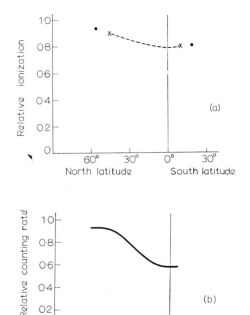

Fig. 7. The latitude effect of cosmic ray intensity. (*a*) Sea-level measurements by J. Clay (1928) using an ion-chamber. The black circles refer to laboratory calibration points, the broken line refers to continuous measurements on board ship. (*b*) Airborne counter-telescope measurements by H. V. Neher at about 10 km altitude, where only about one-third of the atmosphere remains above the apparatus.

apparatus the calibration of which can be relied upon for months on end, and the credit for the first investigation of this effect goes to the distinguished Dutch physicist, J. Clay. In 1928 he mounted a carefully calibrated ion chamber on a ship sailing between Europe and the East Indies, and undertook continuous measurements of cosmic ray intensity on a voyage between these terminal points which,

14

of course, crossed the equator. These measurements and others in the next two or three years indeed showed that a reduction of cosmic ray intensity did occur near the equator (fig. 7 (a)). They were decisively confirmed in very extensive work by A. H. Compton and his co-workers, who set up no fewer than 69 stations equipped with standardized equipment.

The effect shown in fig. 7 (a) may not look dramatic, but it is positive and its correctness is not in question: cosmic ray intensity *is* less near the equator than at higher latitudes. Two reasons for the relatively small effect at sea-level are almost evident. Even if the particles are charged, many may have so much momentum that they will easily reach the earth even at the equator. Moreover particles with a lot of momentum may be much more efficient than those with much less for producing the secondary radiation which is actually measured at sea-level. The measurements support this second view, since the flattening of the curve at latitudes greater than about 40° suggests that additional primary particles arriving at still higher latitudes, and so of even lower momentum, lead to hardly any secondary radiation at sea-level.

The essential conclusion from the work of Clay and of Compton, however, is that some at least of the primary cosmic radiation consists of charged particles moving with momenta great enough to produce effects at sea-level, but nevertheless not too great to be deflected away from the equator. Later (1948) the latitude effect was studied in detail at high altitude in aircraft by H. V. Neher. These flights (fig. 7 (b)) show a very much larger effect, since the effect of atmospheric absorption on secondaries from the lower energy primaries is cut down, and they allow measurements to be made which are precise enough to be used in investigations of anomalies in the earth's magnetic field.

For another important feature let us go back to fig. 6 and consider in particular the path of the particle and the ' earth circle ' which just touch tangentially at the point marked P. These curves represent the situation for a particle in the equatorial plane which is of such a momentum that it can only on a very favourable path reach the earth at all. The particle in question (supposing we are looking down on the earth from above the North Pole) reaches the earth in a very nearly tangential direction from the east. Readers should have no difficulty in satisfying themselves that a particle *of that same sign and that same momentum* cannot possibly reach the same observer horizontally from the west.* This is the basis of the second critical

---

* Try to draw the trajectory of such a particle, bearing in mind that the curvature at P must be of opposite sign as the diagram is seen. Assuming that fig. 6 looks down from above the N geomagnetic pole, for what sign of primary particle is it drawn?

geomagnetic experiment. If the charged particles of primary cosmic radiation are predominantly of one charge or the other, there will be an asymmetry of approach at the equator from east to west, and from that asymmetry the predominant sign of their charge can be determined. If positively and negatively charged particles are present in equal number, there will be no observable effect (and, of course, we must remember that even if there is an effect, any non-charged-particle primary component will dilute what effect there is).

Since no significant number of particles of any sort reach the surface of the earth from near horizontal directions (fig. 4) this is an experiment which has to be carried out at high altitudes, and it was first done by Thomas H. Johnson, an American, with balloon-borne equipment in 1938. His conclusion, which later work has fully supported, was that almost the whole of cosmic radiation falling upon the earth consists of positively charged particles.

Later we shall concentrate on the development of our ideas about the primary radiation: what the particles are, whether there are any negatively charged particles at all, why there is this disparity of charge, what values of momentum exist, where the particles are accelerated, and how they have travelled to the earth. For the moment, however, we are going to assume for simplicity that the primary particles are protons, and in the next chapter tell how we came to understand the development from these of the ' secondary radiation ' which, it seems, is the main feature that is detected at the surface of the earth.

CHAPTER 2
the formation of secondary cosmic rays

2.1. *Primary particles and the atmosphere*

WE begin by considering primary protons reaching the earth: what energy do they have, and what sort of matter is the atmosphere into which they now plunge?

Calculations show that the minimum possible kinetic energy at which a proton can possibly reach points near the equator vertically is about 15 GeV: in temperate latitudes the minimum energy is, of course, lower, say about 1–4 GeV. To understand energies above these minima one must bear in mind that the 'rest mass' of proton 'mc^2' is about 1 GeV, and so all of the particles reaching the earth are 'relativistic'—their kinetic energy is not small compared with the rest energy, and so classical mechanics cannot properly be applied to their motion. Because most readers probably do not know much about relativistic mechanics, we will as far as possible avoid relativistic calculations. The book in this series by W. G. V. Rosser not only covers many general aspects of relativistic physics but also deals specifically with some of the relativistic features of cosmic radiation.

The atmosphere, into which these particles penetrate when they reach the earth, is quite a massive layer of matter, however rarefied it may seem to us moving about in it, corresponding to a layer of mercury about 760 mm thick, or of water about 10 m in depth. Moreover, it is not of uniform density; the density falls away more or less exponentially as we go upwards, decreasing by a factor of 2 for each increase of height of about 5 km. Since it is often necessary to compare what happens in the atmosphere with similar events in layers of condensed matter (water, lead, rock, etc.) a very useful idea is that of 'mass thickness' which describes the thickness of a layer of any sort of matter as the mass of a column of unit area through the layer. The common unit for this purpose is grams per square centimetre (g cm^{-2}).* In these units the thickness of the atmosphere is about 1000 g cm^{-2}; balloons can carry apparatus to within 10 g cm^{-2} of the top of the atmosphere, while at the other extreme, some of the thinnest windows over the apparatus mounted on satellites are as

---

* 1 g cm^{-2}≡10 kg m^{-2}. The former is almost universally used and will be normal in standard texts on these topics for a considerable time. It is often contracted verbally (although not often in print) to 'gram': thus apparatus on a balloon is described as having been exposed for 6 h at 10 g.

little as 100 μg cm^{-2} or even less! The third curve of fig. 2 (p. 5) has, as its main scale, one which measures mass thickness from the top of the atmosphere, reading from right to left.

The atmosphere is thus perceptible for about 50 km above the surface, but large as this extent may seem, it can only be regarded as being thin compared with the distances to which the earth's magnetic field extends at considerable intensity. (At 6000 km above the surface this field has still more than 10% of its value at the surface.)

When a primary cosmic ray proton penetrates into the atmosphere, everything that happens comes about from ' collisions ' of the proton with the matter of air. Such collisions can be of two sorts; they can be collisions either with the electron structure of an atom, or with the atomic nucleus. Since the nucleus is so small the second sort are much less common, but they involve very much greater energy changes, and it is these, when they do happen, which determine the whole form of the secondary cosmic rays.

Collisions of the first sort usually result in an electron being forced out of the structure of the target atom, and are very frequent. This is the mechanism of ' ionization ' in the target material and so these interactions come into their own in our actual detecting devices, ion chambers, Geiger counters and others we shall discuss later. A proton of relativistic energy gives rise to about 60 ionizing collisions in one centimetre of air at normal pressure,* that is to say, in about 1 mg cm^{-2} mass thickness. The energy transferred in each collision is related very much more to the energy of binding of electrons in atoms than to the energy of bombarding particles (providing these have much greater energy) and turns out to vary very little from one substance to another: for many materials it is about 30 eV. Such collisions individually have only the slightest effect on particles such as cosmic ray primaries moving through the atmosphere: any deflection is utterly negligible, while the loss of 30 eV of energy out of many millions is microscopic. But there are a great many of them and the cumulative effect is substantial.

From these figures we may calculate that if a proton were to penetrate down vertically through the whole atmosphere, only interacting with it in this way, then it would lose about 2 GeV of energy in ionizing the air. For this reason, even if there were no others, a primary cosmic ray particle initially of less than 2 GeV kinetic energy when it reaches the earth's atmosphere cannot at first sight be expected to produce any observable effect at sea-level.

---

* Strictly, this proton will itself make about 30 ionizing collisions in 1 cm air, but some of the ejected electrons are of sufficient kinetic energy to produce further acts of ionization. Thus about 30 acts of primary ionization lead to roughly an equal number of secondary ionization, and so perhaps 60 in all.

The diagrams in fig. 2 were intended to show the expansion of knowledge in the early days of cosmic ray studies as it became possible to use apparatus at increasingly greater distances from the surface of the earth. Fig. 8 shows how this curve has been extended now that apparatus can be sent to very much greater distances using rockets and satellites. It shows again, and even more convincingly, how the primary radiation, which we have now also found out to be mostly charged particles (perhaps protons), actually leads to secondary effects

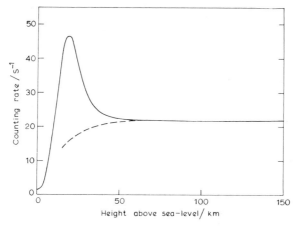

Fig. 8. Counting rate of a single Geiger counter as a function of altitude obtained in an early rocket experiment. Notice in particular the steady level of the count due to primary particles until secondary production begins at an altitude of about 60 km. The broken line is referred to in the text (p. 19).

in the atmosphere. What it makes clear is that this is not the effect which would be observed if the ionization process, which we have just described, were an important one. If it were, the effect would be more like that shown in the broken line, a decrease of intensity taking place as the protons of low energy use up the whole of their kinetic energy in ionizing the atmosphere, and so reach the end of their ' range '. The only way in which *what is observed* can have happened is by events taking place which lead to an *increase* in the number of ionizing particles present—that is to say, by the development of a ' secondary radiation ' in a way which must involve the production of many secondary particles from a single primary—and so, in extreme examples, leading to the formation of the ' showers ' we have already mentioned. In fact, the curves shown in fig. 2 (*c*) and fig. 8 are to all intents and purposes Rossi curves (fig. 5) in the atmosphere instead of in lead.

19

2.2. *Nuclear collisions*

To identify what is happening when these showers are initiated, we have to develop ideas about nuclear collisions and, first of all, to notice that what are important are not the rather distant collisions of nuclei which lead to 'Rutherford scattering' but the close collisions which lead to nuclear disintegrations. It is only in such collisions that we can expect to find ways in which the number of charged particles present can be increased.

To get some idea about the possibility of such collisions in the atmosphere, we start with the 'geometrical size' of a nucleus, that is to say the size into which the particles forming the nucleus seem to be packed: this corresponds to a 'target area' of about 10^{-28} m². Most matter contains about 10^{22} atoms per gram,* therefore close nuclear collisions are likely on average to happen about once in 100 g cm^{-2}; for air the required thickness will be rather less (in fact ~ 80 g cm^{-2}) while in lead it will be somewhat greater.

So a proton coming into the earth's atmosphere is likely to make its first collision high in the air, and it has an almost negligible chance of penetrating right down to the ground without making any disintegrating collision at all.

The study of nuclear collisions by cosmic ray particles has had a particularly important part in deciding the way physics as a whole developed in the last twenty years. It has let us construct a useful description of secondary cosmic radiation so that we can interpret what we observe in terms of the primary radiation from which it has been derived. However it also, at a very early stage, showed quite unexpected features about the interactions of nuclei and about nuclear particles which are so interesting that large machines were quickly built in Western Europe, the United States and in the U.S.S.R. to study them in detail under laboratory conditions rather than in the far more difficult conditions of natural cosmic rays. The details of these studies have now come to be a dominant part of nuclear physics —'high energy nuclear physics'—which brings in many complex new ideas. Here we need to make use of only some of the most straightforward features; this will save the reader from confusion, and can be excused of the authors since anything more detailed is not at all well understood by anybody in relation to the very high energies concerned in the secondary cosmic radiation!

The open-ended nature of modern high energy nuclear physics, which has grown from the features of cosmic rays which we are about

---

* Using the Avogadro number, $N_A = 6 \times 10^{23}$ atoms per mole, we have the following values:

Material	Air	Water	Iron	Lead
Atoms per gram	2×10^{22}	3×10^{22}	10^{22}	3×10^{21}

to discuss, has by now become embarrassing, for there seems to be no limit to the size and expense of the very powerful machines now used, on every occasion with some confidence that new complexities will be revealed.

To learn about the collisions of primary cosmic ray particles it is really necessary to take apparatus to the height at which such collisions are commonest: this means to a height of perhaps 30 km. In fact the first steps were taken at much lower altitudes, at mountain observatories, and the data coming out of this work are about phenomena within the development of the secondary radiation. The assumption that secondary protons would behave in exactly the same way as primary protons of the same energy has never been seriously questioned. Although these experiments provided a basis of understanding which was pretty well correct, the extension to the optimum height was important, partly to provide much more material but also to see the nature of the processes was not different for particles which were certainly primary and often much more energetic. Even now the second objective, that of understanding the important features of collisions at the highest energies is something about which there is a great deal of uncertainty.

2.3. Nuclear emulsions

The tool which was to dominate the early work on particle collisions was developed by Cecil Powell, and the team of enthusiastic young men who joined him at Bristol. It had been known for almost fifty years that ionizing radiations affected photographic emulsions and produced developable areas on plates or films as does light. Moreover, there were indications that the ionization due to a single particle could leave something like a ' track ' in the emulsion, although these ' tracks ' were hopelessly crude compared with those in a cloud chamber, and not of any evident use. Powell set himself the task of developing the most effective emulsions for recording the passage of ionizing particles, and in this work he had help from major firms producing photographic materials. All existing emulsions had been optimized for special purposes—for normal photography, for X-ray purposes, even for infra-red photography—and there was no reason at all to think that any of them would by chance be best for Powell's purpose. They were not, and special nuclear emulsions were developed which transformed the situation. The crude hints of tracks became thin, sharp lines in which not only the exact path followed by particles became clear, including the effects of deflections (sometimes described as scattering) which occur as the particles passed near to atomic nuclei, but also the intensity of ionization along each track could be measured.

Such emulsions were ideal for exposure in balloon experiments: they formed a static packet which had only to be taken up and got down again (and recovered when it had come down): there were no complications of machinery or electronics which even as recently as 1950 would have made the whole venture more liable to failure and frustration.

As often happens the straightforward and basic discoveries which this new technique made possible came quite quickly, but experiments using nuclear emulsions, much more refined and more specialized, are still proving valuable at the present time. What came out of the early experiments related to two distinct problems. For the first time observations were made which related to individual primary particles and from which information could be derived as to what they really were (we have up to now found it simplest to think of them as protons), and further, the nuclear collisions of these particles could be studied, not indeed those with the nuclei of the air, but rather those with nuclei of silver and of bromine in the emulsion. It was found that the primary radiation was (as far as these experiments went) entirely made up of positively charged particles which were in fact atomic nuclei. Protons were the commonest, but other nuclei were undoubtedly there and particularly α-particles. Later, we return to this question of the composition of the primary radiation, but our main interest at this stage is in the products of the nuclear collision, and here we do not lose very much if we go on thinking of the primary material as protons, for in secondary formation, most complicated nuclei act mainly as groups of protons and neutrons which act almost independently.

A nuclear collision in (say) oxygen by a proton of several GeV energy usually leaves the incident proton with a considerable fraction (perhaps as much as one half) of its original energy, so, since it is still very high in the atmosphere, it can repeat the performance several times over before it has lost so much energy as to be very different in its behaviour. On each occasion it will be accompanied by several protons and neutrons detached from the target nucleus. The number of these will vary from event to event, one obvious reason for this is that some collisions will be nearly head-on while others will be more or less grazing. These extra nucleons* will be of distinctly lower energy than the primary particle, for all their kinetic energy has come from it, but from a really energetic primary particle they may very well have all the potentialities of actual primary particles of lower energy. Thus in a succession of nuclear collisions, the primary

---

* Protons and neutrons differ in the very noticeable property of charge, but in all other ways are found to be extremely similar. In the nuclear collisions dealt with their charge is not an important property and we need a collective word to describe protons and neutrons when their similarity is the important feature rather than their difference. Then we describe them as ' nucleons '.

protons, although now significantly less energetic, will be found accompanied by a retinue of other nucleons which have, as it were, been pressed into their company. This group structure of nucleons travelling together but with their total kinetic energy entirely dependent on what was brought in by a single primary particle is described as a ' nucleon cascade '. The particles are not travelling along strictly in the same direction, they will be found to be spreading out slightly, often by angles much less than a degree. It is only within the ' nucleon cascade ' that further nuclear collisions take place.

2.4. Meson formation

Now the most important feature of the whole operation remains to be described. Nuclear interactions are found not to be elastic: that is to say, the energy brought in by the incoming proton is not all accounted for by the energy of the outgoing particles of the nucleon cascade. Some energy has gone somewhere else, and it has in fact gone towards the creation of entirely new particles and supplying the kinetic energy with which they come away from the point of inter-action. It is these new particles, and what becomes of them, which determine almost everything which we observe about secondary cosmic radiation. Studies in high energy nuclear physics have shown that very many different kinds of new particle can be created in nuclear collisions, and the study of them, and especially the attempts to put them into some sort of orderly scheme, has been one of the great objectives of high-energy physics at the large accelerators over the last 20 years. However, for a general understanding of secondary cosmic radiation we need only consider the commonest sort of new particle, which was identified in the classical emulsion work of Cecil Powell. This particle is called the ' pion ' (symbol π) and the emission of pions in a collision between nucleons has a great deal in common with the emission of light quanta in collisions between charged particles, for example in the emission of X-rays in an ordinary X-ray tube. It was one of the triumphs of the theoretical physics of particles that the general nature of pions should have been predicted following this line of argument before they had ever been observed. In particular, there were predictions which turned out to be sub-stantially correct about the mass of the particles and about their instability. It is a characteristic of all the new particles created in nuclear interactions that they should be unstable, and when, in one or more steps, they reach a stable configuration, this is not surprisingly found to comprise only the familiar and stable particles of matter!

Pions are found to exist in three states of charge, they have either one positive or one negative electronic charge, or no charge at all, and the history of their instability is set out in the following three equations:

$$\pi^+ \rightarrow \mu^+ + \nu$$
$$\pi^- \rightarrow \mu^- + \nu$$
$$\pi^0 \rightarrow \gamma + \gamma.$$

From these equations we can see that the two charged forms of the pion behave similarly. Each decays in a time of about 0·02 microsecond to form another hitherto unknown unstable particle, the muon (symbol μ). This in turn decays to produce an electron, but the life-time of its transformation is very much longer, about 2 microseconds, and in both transformations some energy goes to another elusive and practically undetectable particle, perhaps unfamiliar, the neutrino* (symbol ν). We write, for muon decay:

$$\mu^+ \rightarrow e^+ + \nu + \nu$$
$$\mu^- \rightarrow e^- + \nu + \nu.$$

This is very surprising behaviour, and at first sight, in spite of predictions about the existence of the pion, difficult to accept as an important feature in the development of physics so it is worth while spending a little time in putting these particles and their properties into context.

First of all, they are outside our general experience because of their very short life-time relative to the time-scale of our normal activities, and it was thus inevitable that they should remain unknown until experiments were possible on a time-scale much shorter than could be based on the ordinary use of our unaided senses. In fact, one group of particles, the muons, do live for quite a long time (two millionths of a second!) and the majority of single particles of cosmic radiation observed at the surface of the earth are indeed muons. It was when these were studied in detail that it first became apparent that particles occurred in cosmic rays with properties definitely different from those of known particles and it became clear in laboratory experiments not only that these particles were unstable, for the very act of decay into an electron could be photographed in a cloud chamber, but also that they were about 200 times as heavy as electrons.

The new unstable particles, like all other matter, are subject to the ordinary general rules of physics and in particular obey the conservation laws. The modes of decay which are set out in the equations above are possible changes which can take place within the framework of conservation of energy, of momentum and, of course, charge. In general it seems that if such changes can take place on

---

* The neutrinos emitted in pion decay differ from those from muon decay, but for our present discussion this difference is not important. Nor is that between ' particles ' and ' antiparticles ' (including neutrinos and antineutrinos) which would appear in a more complete formulation of these processes.

these terms they will do so, although sometimes very slowly; if they cannot, then the particle is a stable one and thus one of the more well-known particles of physics.

Regarding the development of secondary cosmic radiation in the atmosphere, we see that charged pions lead successively to muons and electrons in two stages of decay, but as is indicated in the equations, uncharged pions decay in a totally different way. The product of

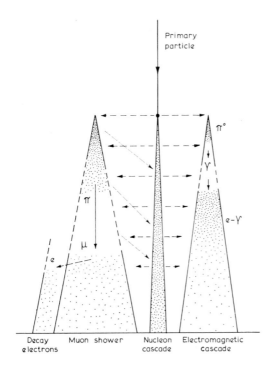

Fig. 9. Shower development in the atmosphere: the various components, although in reality superposed have been separated here for clarity. The arrows indicate the transfer of energy within and between the components.

decay for each neutral pion consists of two quanta of radiation, and the lifetime before this takes place is very short indeed, less than one millionth of that of the charged pions. Accordingly, the charged pions on the one hand, and the uncharged on the other lead to two completely separate secondary cosmic ray components, and in fig. 9

25

these are shown in a diagrammatic way alongside the third component which we have already discussed, the nucleon cascade. Here it has seemed easiest to shift the two pion components, one to the right and one to the left, so that the development of the three components can be seen separately, but these three systems of course actually develop one on top of the other.

The decay of charged pions leads to positive and negative muons and these have lifetimes of about two microseconds before they decay to yield ordinary electrons. Now light travels only about 600 m in two microseconds and here we are thinking about particles created perhaps 10 km above the surface of the earth. At first sight it would seem certain that all of the muons would decay high in the atmosphere, and what would remain would be the electrons which represent the final stage of the pion–muon–electron series of transformations. Indeed, quite a large number of muons do decay in this way and the electrons produced behave as other extremely energetic electrons do, but many muons, and particularly the more energetic ones, do not decay before they reach the ground. The reason for this apparently unlikely behaviour is to be found in one of the features of the special theory of relativity according to which we have to make quite clear, in considering the behaviour of a particle which is apparently moving with a velocity close to that of light as seen by a particular observer, for example by ourselves, whether we are timing events concerning it with our own clocks or with imaginary clocks moving with the particle itself. If it were possible for a clock to travel with the muon, then muon decay would on average be found to take place when that clock had registered about 2 microseconds; such a clock is the appropriate one for timing the internal history of the muon, and, indeed, the probability of this decay is itself precisely such a clock. However, if apparatus capable of measuring distance travelled also accompanied the muon, the distance from the point of interaction to the surface of the earth would be measured as something very much shorter than that deduced by an earth-bound observer. This is the famous so-called ' Lorentz contraction '. An observer travelling with the muon would naturally regard his own clock as correct and if he was able to see a clock upon the earth would regard it as going very slowly indeed. He would equally regard his measurement of distance correct and he would come to the conclusion that the earth-bound observer, who was under the impression that the muon had travelled 10 kilometres, must be using a measuring rod which was somehow far shorter than its reputed length.

The phenomena of the Lorentz contraction are described in terms of a constant, ' gamma ' (γ), which describes the amount by which the earthbound measuring rod appears contracted, and the amount by which the clock, either of the earth-bound observer seen from the

26

muon or the muon's clock seen from the earth, appears to be going slow. The fundamental equations are:

$$\gamma = (1 - v^2/c^2)^{-1/2}$$
$$E = \gamma mc^2$$
$$(\Delta x)' = \gamma(\Delta x), \quad (\Delta t)' = \gamma(\Delta t),$$

where m is the mass of the muon measured in its own frame of reference (the 'rest-mass'), v is the relative velocity of observer and muon (assumed approaching in the line of sight), E the total energy of the muon, c is the velocity of light, (Δx), (Δt) intervals of distance travelled and of time measured with instruments accompanying the muon and $(\Delta x)'$, $(\Delta t)'$, the corresponding observations of an observer on the earth*. The important point to notice here is that the quantity γ does not become insensitive to further changes of energy as the velocity of the particle is approaching that of light, but goes on increasing indefinitely as the energy of the particle increases, in spite of the fact that it has reached a velocity very close indeed to the velocity of light. As an example, a muon of energy 10 GeV has a value of γ of about 100 (the rest-mass of a muon being approximately 100 MeV) and, according to an earth-bound clock, it can therefore travel for about 200 μs before it is likely to decay, that is to say for a distance of something like 50 or 60 km.

We have had to go into a good deal of detail about this particular picture to show how there is nothing surprising in the fact that muons from collisions high in the atmosphere are found to reach the earth before they decay, so on the left hand side of our diagram (fig. 9), which shows the shower component arising from charged pions, we represent a few muons decaying to yield electrons but the majority going on and reaching the earth unchanged. That this actually takes place brings out another property of the muon: it is not subject to absorbing effects in nuclear collisions, for in passing through the whole atmosphere it must, as a rule, pass right through the nuclei of several atoms on the way and it seems to do this without any significant effect. This is a property in which nucleons (protons and neutrons) differ completely from particles such as muons and electrons. No proton or neutron can pass through a nucleus unaffected, electrons and muons almost always do. This distinction is brought-out by describing neutrons and protons as 'strongly-interacting' particles, and electrons and muons as 'weakly-inter-acting' (or for most purposes non-interacting) particles. In this scheme of things, the pion turns out to be strongly-interacting, and therefore our diagram shows a return arrow to the nuclear cascade

---

* This particular situation is treated in detail in W. G. V. Rosser's book *Relativity and High Energy Physics* [No. 7 in the Wykeham Science Series] (pp. 76–78).

from the pion component, since there is a possibility (sometimes quite a strong one) that the pions, which generate the muon shower in the left-hand sector of the diagram, may, before they have time to decay, be involved in nuclear collisions from which both the nucleon cascade and also the two pion components are further reinforced. We have to note this complication but it is hardly worth pursuing further.

The situation regarding the products of decay of neutral pions is quite different. These are from the outset familiar and normal particles, photons, and what is more, the lifetime for decay of neutral pions is so extremely short that no considerations of the Lorentz transformations can stretch out the life of a neutral pion as seen from the earth beyond the time which it takes to travel something less than a single centimetre. Consequently, the possibility of nuclear inter-action preceding decay can be ignored. So right at the outset, and high in the atmosphere, these pions produced in nuclear interactions yield light quanta of very high energy, in fact of energy not a great deal lower than the prevailing energies of nucleons in the same region.

The behaviour of energetic quanta or γ-rays when they pass through matter arises from three or four possible and quite well known processes, the importance of which vary with the energy of the quanta. At a certain energy one process may be completely dominant, at another some other process will be the important one. But at the extremely high energies we are now discussing, only one process is of any importance at all, this is the so-called ' pair-production ' process in which a γ-ray gives up the whole of its energy to produce an electron pair, that is for the creation of a negative electron and a positive electron:

$$\gamma \rightarrow e^+ + e^-$$

$$h\nu = 2mc^2 + \text{kinetic energy,}$$

where the first line gives the particle description and the second the energy balance. This process is known (although it is then only of minor importance) at much lower energies, starting at a threshold energy, as will be obvious from the above equation, when the γ-ray has just enough energy to provide for the rest-mass of the two created electrons. This is actually about 1 MeV, and so the harder γ-rays from radioactive sources (for example those from ^{60}Co) yield a certain amount of pair production in passage through matter.

Pair production necessarily depends upon the possibility of the reaction taking place within the framework of the conservation laws (and we make no apology for stressing once again so fundamental a controlling feature of all physical behaviour), and this transformation is not possible in a vacuum*. There must be interaction with another body that can take up some momentum, and this interaction is normally through the electromagnetic interaction with the field of an atomic nucleus: pair production is only possible for γ-rays when they traverse matter.

Because the momentum transfer in the γ-electron system is not fixed, neither is the angle between the trajectories of the two created electrons, but from general considerations of relativity, the average angle of separation decreases with increasing energy, and for all effects discussed in this book it will be accurate enough to think of both the electrons as moving off more or less in the direction of motion of the γ-quantum from which they have been derived.

We have now to consider in turn what happens when extremely energetic electrons, such as those created in cosmic ray pair-production, pass through matter. For these the picture is very much as it was for energetic γ-rays: of several processes by which electrons are known to lose energy in passing through matter one is quite dominant at these extremely high energies, and this consists of the emission of γ-rays, often individually taking a substantial fraction of the whole energy of the emitting electron. These emissions take place as the electron is suddenly accelerated in the electric field of a nucleus when it passes close to it, and in this way it has close similarity with the emission of X-rays in an X-ray tube when the accelerated electrons are suddenly brought to rest in the material of the anti-cathode. Rather strangely, we have no English word to describe this particular emission process, and we have adopted a name from German: it is described as ' bremsstrahlung ' emission.

2.5. *Growth and decline of showers*

One now sees that here there is a sort of repetitive process, and the initial steps are illustrated in fig. 10. In this very high energy region both γ-rays and electrons interact with the matter they are passing through almost entirely by complementary processes: the γ-rays yield electron pairs, these produce more γ-rays and so on. Thus from a single uncharged pion, we rapidly develop a quite complex system of electrons and γ-rays and this complex is known as the electron-photon cascade, or sometimes the electromagnetic cascade. Like the whole of the development from the initial primary entering the atmosphere, this section exists solely on the energy deriving from it, here that emitted in nuclear interactions as neutral pions. The

* It is a useful exercise for readers to satisfy themselves that pair production in a vacuum is indeed excluded by the conservation laws. The momentum of a quantum of energy $h\nu$ is $h\nu/c$, while the total energy, rest mass and momentum, p, of a particle of mass m (here an electron) are related by the expression $E^2 = (mc^2)^2 + (pc)^2$.

Fig. 10. The initial steps of an electromagnetic cascade. In a nuclear
interaction, one of the products is a neutral pion (thick broken line).
This decays to yield two photons (thin broken lines) and each of these
interacts to yield an electron pair (full lines). The resulting electrons
emit bremsstrahlung photons which in turn yield further electron pairs.
(The other, charged, products of the initial nuclear interaction are not
electrons, but are also shown as full lines.)

successive stages increase the number of electrons and photons, but
the total amount of energy does not increase and so, as the number of
particles grows larger, the average energy of each becomes smaller.
This process goes on until the conditions under which we have described the development, that is to say, the conditions that pair
productions and γ-ray emission are the only processes of interaction
that matter, cease to be true. When this stage is reached, particles
and photons also lose energy by other processes well known in nuclear
physics, such as ionization and the Compton effect, which do not
involve the multiplication of the number of particles, but which do
abstract energy, and so the shower complex of electrons and photons
which had been growing in numbers begins to be absorbed. Eventually
all the initial energy will have gone in processes such as ionization,
and so ultimately into a slight heating of the atmosphere where this
has taken place.

Returning to fig. 9, we can now see how it is that the development of the secondary component of cosmic rays when a primary particle enters the atmosphere can properly be set out as a superposition of three components, the nuclear cascade, the muon component and the electron–photon cascade. All of these exist, as it were, one on top of the other and mixed together. Moreover, at each important nuclear interaction within the nuclear cascade, further contributions feed into the muon distribution and into electron–photon cascades. In fact, the nucleon cascade renews these two other components all the way down until it is expended. Each act of particle formation involves some spreading from the originally defined direction and these particles in traversing the air may be further spread by scattering. The degree of spreading differs for the various components. It is greatest for the muon component and least for the nucleonic component, with the electron–photon cascades spreading to an intermediate degree.

If we now think about the whole complex of the secondary component from a single primary particle, and follow it down through the atmosphere, it will be evident that the more energetic the primary particle the longer it will be before the various components pass over from the growing stage to the absorption phase. For primaries of lowish energy, say 10 GeV, the probability of anything at all getting down to sea-level is pretty small. But if anything does, then it will most probably be only a single muon derived from a single charged pion which in the very first collision was formed with an exceptionally large fraction of the whole available energy. At a primary energy of say 100 GeV it is fairly certain that several muons will reach sea-level, although the nucleonic component and the electron–photon cascade are likely to have been completely absorbed well above the surface of the earth. These few muons, however, will be very much spread out and may easily be a hundred metres apart when they reach the ground, so it is not likely that an observer will establish that any two of them are connected; he probably will not have apparatus which will allow him to relate any two of them* and so these muons will all be observed, if they are observed at all, as though they were single

* It is interesting to carry out a simple calculation to decide whether this is even possible. The rate of arrival of muons at sea-level is about 100 m^{-2} s^{-1}, and it would be reasonable to consider all other muons falling within a radius of 100 metres from the arrival point of our particular muon. One can assume that a related muon (from the same primary interaction) would arrive at exactly the same time apart from the fact that they may not be coming exactly vertically but at an angle θ (up to say 30°) from the vertical. For non-vertical *related* muons there will be a time-difference in reaching level ground, since one may have to go up to 100 sin θ metres further than the other. Compare the time corresponding to this difference of path with the average time-separation of totally unrelated muons falling within the 100 m circle.

isolated particles. The vast majority of cosmic ray particles detected at ground level are thus seen, as it were, singly and as though they were not members of a complicated shower development of the sort we have been discussing. Most of these are muons and the actual number of them surprisingly large; several thousand pass through an average living room every second.

It is only when primary energies of perhaps 10 TeV (i.e. 10^{13} eV) and more are involved that the situation becomes different, for it is not until there is some likelihood of the electron–photon cascade reaching the ground that one can expect easily to be able to observe several particles at once, and to make such observations as those involved in Rossi's experiment (p. 9). We now can identify this notable piece of work as an experiment on the electron–photon cascade of a primary particle energetic enough for this component to reach down to the ground, and, what is more, to reach the ground with relatively energetic electrons and photons present, and so with the electron–photon cascade in the growth state. In a thin layer of solid matter (fig. 5) such a cascade will show an increase in the number of particles present, and some cascades which without the extra layer would have had little chance of discharging all of Rossi's counters, then, with more particles in them, are able to do so. It is in this way that the Rossi increase of counting rate occurs. But in a thick condensed layer of matter such as lead the end of the growth phase is soon reached, and the later part of the Rossi curve therefore shows the absorption phase of the electron–photon cascades, and then finally a ' tail ' caused by quite new cascades initiated by what is left of the nucleon cascade is what remains.

It is from such energies upwards that primary cosmic ray particles yield the characteristic features of a shower, and, in effect, any arrangement of detectors which detects showers is a device for detecting the products secondary in the atmosphere from primary particles of far above average energy. No two showers are exactly alike, for every step in the development, the distance, for example, that a γ-ray travels before initiating pair-production, involves a random element, and what can be calculated has to be based upon the average distance before this single act occurs, or more elaborately, the probability that it will happen at a particular distance. Descriptions of showers are always statistical in nature, and these too either describe an ' average shower ' from a particular primary, or attempt to give also some indication of the fluctuations from this average which can be expected.

2.6. *Secondary cosmic rays in extra-terrestrial studies*

We have discussed at rather great length the development of secondary cosmic ray radiation, and it is as well to bear in mind that we have been doing this not so much for its own sake, interesting though the processes are, and important as the initial observations of unstable particles quite certainly were, but mainly because of the extent to which one can use observations upon secondary particles to study primary phenomena. If one is going to observe primary particles, the experiments are costly even when they are possible, and insofar as the same information can be got conveniently from ground-based observations of the secondary component this is very often to be preferred, while if observations on some primaries are not possible at all, deductions from the secondary radiation will of course be our only source of information.

From this point of view our conclusions can be summarized as follows:

1. For a primary of low energy the likelihood of any secondary particles reaching the earth is low, hence direct observations at balloon altitude or on satellites are a necessity. However, the intensity of the primary beam, that is to say, the number of particles striking unit area in unit time, is quite high at these energies, and the sort of dimensions that can be accommodated in a satellite experiment are adequate to detect these particles directly in large enough quantities and to make observations upon them.

2. For primaries of higher energies, the flux of particles quickly goes down (see for example fig. 26) and eventually, when we are thinking of primary energies such as 10^{19} eV, experiments show that of these particles only perhaps three or four reach the earth per square kilometre per year! Here direct experiments on the primary particles cease to be possible, for any conceivable apparatus, whether on a balloon or on a satellite, cannot possibly have an area anything like 1 km², nor are we prepared to support it for a great many years in the hope that ultimately one such primary particle will be observed. Here we are forced, whether we like it or not, to get what information we can from the secondary cosmic radiation and in many ways the development of secondary radiation through the atmosphere could hardly be better for this purpose. For such extremely energetic primary particles the electron–photon cascade is near the maximum of its development close to ground level, and in place of a single primary we have perhaps a thousand million or more secondary electrons, photons and muons spread out over several square kilometres of the surface of the earth. Such a shower can be detected all over this area and useful measurements upon it are, as we shall see, possible even when no detector lies in the direct line of the path of the primary particle. It is in this way that we are able to make any measurements at all about the exceptionally energetic primary particles. The information is not as exact or as complete as one could imagine from an idealized experiment made directly on the

primary particle, but such an idealized experiment could not be translated into reality and so one here deals with the real, if not perfect, experiment that you can do rather than the ideal experiment that one cannot possibly do.

3. There is, of course, an intermediate region in which direct observations on primary particles become increasingly hard to carry out, as the energy of the primaries increase and so their flux decreases, and where the choice of observations on primaries or secondaries will depend on the objective of a particular piece of work. Good observations on primary particles are by their very nature better than corresponding observations on secondary particles, if only because the chance element in each step of the secondary development (the actual distance down through the atmosphere which the primary penetrates before the first collision takes place, the distribution of energy in any single nuclear interaction between the leading particle, other nucleons, charged pions and neutral pions and so on) smears out the characteristics of the primary. Because of this succession of chance variations, primaries of the same energy but different mass (and charge) may produce almost indistinguishable secondary showers, and equally, primaries of energies differing by a factor of two or more might in many detector systems yield indistinguishable records of the secondary showers.

It will be enough to quote two extreme examples to illustrate the conditions which determine experiments in this intermediate zone.

(i) Developments of the emulsion technique at Bristol and elsewhere, in which very large stacks of emulsion, sometimes with absorbing layers interleaved in them, have been flown for many hours at the highest possible altitudes, have provided a few examples of primary particles which can be identified as of mass as great as, and perhaps greater than any stable element on earth, and also a small number of records of the initial stages of secondary component development from primary particles up to energies like 10^2 TeV (10^{14} eV) in great detail. No information at all on these features could possibly be derived from ground-based experiments, yet at altitudes where the primary particles are accessible, valuable information comes from even two or three precise and detailed individual observations.

(ii) In experiments to determine and interpret time-variations and direction-variations of the primary flux, even at much lower energies, the numbers which can be collected in above-the-atmosphere experiments are still quite inadequate to provide the statistical accuracy which we need in experiments where variations of much less than 1% may have to be established to a high degree of probability. Moreover, some time-variations have a period of several years! Under these conditions, ground-based experiments become those which will certainly prove more effective: it is assumed in carrying them out

34

(and almost certainly correctly) that the uncertainties about the nature and energy of the primary particles which are introduced by making use of secondary data, and which arise in ways we have just indicated, are only of minor importance, and that in spite of them, variations observed exclusively on the secondary component provide valuable data.

2.7. Atmospheric variations

The use of observations on the secondary component to monitor changes in the primary flux must take account of the fact that the atmosphere does not constitute an unchanging layer of matter, and data for this purpose require corrections to be applied to reduce actual measurements to values which would be recorded at some standard atmospheric condition. These corrections are quite complicated: they depend on such features as the barometric pressure (perhaps the most obvious, since it describes the atmosphere in terms of mass-thickness of material) and the temperature distribution throughout the atmosphere and it will be evident from what has already been described that these corrections will vary from one component of the secondary radiation to another. To add to the problem many detector systems are sensitive to more than one of the components incident on them, and under these conditions it is often necessary to determine the characteristics of a particular piece of apparatus with respect to these changes of the atmosphere. Here we will refer to only two variations—those arising from changes of pressure and of the general temperature of the atmosphere.

The pressure correction for the ordinary cosmic ray intensity, which is almost entirely made up of individual muons and electrons, is not large: the muons themselves are very penetrating and most of the electrons involved are decay-products of muons. The effect is of the order -0.35% per mm of mercury (the negative sign indicating that an increase of pressure leads to a decrease of measured cosmic ray intensity). Since measurements of variations of this component are mostly used for determining very long-term averages, an effect of this magnitude does not present great difficulties.

The corresponding correction for measurements of the nucleonic component is larger, about -1% per mm of mercury and since this component is used for studies of rather rapid changes of the low energy primary flux, it has to be measured with greater precision. Fortunately, for the nucleonic component other atmospheric variations have very slight effects, and the most difficult element in correcting these measurements rather surprisingly lies in our ability to pick up the true, ' mass thickness ', barometric pressure at the detecting equipment against the dynamic pressure arising from wind conditions. Even when every effort is made to obtain the true static

35

pressure, reliability and steadiness of the pressure records is very difficult to achieve in very windy weather, and the recording barometer which can be seen in fig. 13 (*b*) is connected in a quite complicated way with the external air.

A negative correlation between the ground temperature and the general cosmic ray intensity was related, as long ago as 1938, by P. M. S. Blackett to the decay of muons. This effect, about -0.3% K^{-1} arises because at higher temperatures the whole scale of the atmosphere is extended and muons have longer paths to travel and in which to decay. More therefore do so. A close study of this effect needs information about atmospheric temperature not only at ground level but at all heights up to the main regions of origin of muons. It will be evident that a variety of secondary ' variations ', for example a seasonal variation, stem from this effect.

CHAPTER 3
solar modulation

3.1. *Identification of solar effects*

INCOMING primary particles at the earth can be described by their nature, their energy, their direction of arrival (either in simple terrestrial co-ordinates or, corrected for the influence of the earth if necessary, in galactic or in astronomical co-ordinates) and their time of arrival, and it would be all too easy, but quite useless, to make nothing more of this than a long catalogue of observations. What is needed is something much more structured, to relate the observations in the first place to phenomena we already understand and so to lead on to new deductions about what has been unfamiliar. Very early in the book we used just this approach to deduce that the primary cosmic rays were mostly positively charged particles. In a similar way, we now examine a special category of variations of cosmic ray intensity, in arrival direction and in time, which we think are features of the solar system. These are broadly effects which originate at the sun but which spread out from it certainly far beyond the orbit of the earth and even perhaps beyond the distant planets.

We have first to explain how and why one can hope to distinguish solar effects from variations which might have been present in the radiation long before it reached the solar system. To do this we appeal to two features, the argument from either of which is almost conclusive.

(i) Any time-variation of cosmic rays from outside the solar system will be 'slow' compared with our normal time-scale, and of simple form. We shall exemplify this feature in more detail: briefly it is because the particles have come by extremely long paths which represent diffusive movement from independent, very distant, source regions. Without attempting calculations, the sort of variations which might arise from the variation of these sources would probably always need at least centuries of measurement to detect rather than days, months or years. On the other hand, we know very well that solar phenomena of other kinds, for example magnetic storms and effects leading to interference with radio communications, can happen quickly sometimes giving effects which are measurable within minutes. The likelihood that quick changes of cosmic ray intensity are solar-related therefore seems well-based, and it becomes a near-certainty when it turns out that often cosmic ray changes can be directly related to these other phenomena which are known to be of solar origin.

(ii) All observations about variability, either in time or direction, are further modified because we make them from the surface of the spinning earth: our equipment scans around us as the earth rotates. Now the apparent rotation of the earth relative to the sun is not the same as its rotation relative to very distant objects (that which used to be described as ' relative to the fixed stars '). For everyday purposes and for obvious reasons we use solar time, but the astronomers, except when they are specially concerned with solar phenomena, use ' sidereal ' time, time based on the rotation of the earth relative to distant stars and galaxies. Solar time and sidereal time differ by one part in 365, for the difference arises in a cycle of one year because of the orbital motion of the earth about the sun.

As a consequence, *directional* features of cosmic ray intensity which are more or less steady over a long period of time will be observed as a *time-variation* by the scanning apparatus on the earth: if the directional variation is something which is built up in the geometry of the solar system, this scanning time-variation will be repeatable in solar time: if on the other hand, it is an effect of interstellar cosmic radiation, and was there even before the radiation came to the solar system, the time-variation will be repeatable in sidereal time. Over two or three days, the difference between solar time and sidereal time will not be noticeable, but when measurements go on for a large part of a year or, better, for years on end, a sidereal variation can be distinguished with certainty from one in solar time.

Prolonged analysis in sidereal time has established limits within which we are confident that we can regard the flux of cosmic radiation coming to the solar system from outside as isotropic (that is to say, coming equally from all directions), and it is against this background that we pick out solar features. For any particular energy of cosmic ray particles we can say definitely that all the solar effects which we are now going to discuss are a great deal larger than any sidereal variation against which they have to be picked up—to all intents and purposes for what now follows the incoming radiation to the solar system can be taken to be strictly isotropic.

The solar effects are of two kinds:

(i) The sun itself is a very weak and intermittent source of cosmic-ray-like particles, so intermittent that noticeable effects at the surface of the earth, which last only for a few hours, occur less often than once a year. The particles coming from the sun on these occasions are of much lower average energy than are the main flux of particles from outside.

(ii) Conditions within the solar system are continuously affecting the incoming cosmic ray flux, and the various changes arising from this cause are described as 'modulation' of the galactic cosmic rays. We shall concentrate upon the second of these, the modulation effects, for the emission of cosmic ray particles from the sun provides relatively little data and, moreover, has been very slight indeed during the last ten years or so during which, from all points of view, the sun has been unusually 'quiet'. By contrast, the modulation effects yield continuous data. The three main effects, which are interrelated, are:

    (i) a solar diurnal effect,
    (ii) the Forbush effect,
    (iii) the 11-year variation of intensity.

All of these, and also the corresponding transmission effects for particles originating at the sun, are related to properties of inter-planetary space, and to the 'solar wind' which blows in it.

3.2. The solar wind

The 'solar wind' is the normal state of the sun's atmosphere. Any atmosphere, whether it is that of the earth, the sun or anything else has to be heated if it is not to condense completely. The nature of the heating determines what sort of atmosphere is produced. The terrestrial atmosphere is heated almost entirely from outside, by solar radiation. A great deal of this is absorbed in the upper atmosphere while some penetrates and heats the solid or liquid surface of earth. The result is an inner or lower atmosphere heated almost entirely from below so that gentle convection takes place in the so-called troposphere, and an outer (upper) atmosphere, partly ionized and mostly heated at the top, and thus attaining a temperature gradient which stabilizes it, leading to the stratosphere and the ionosphere. The supply of anything which could escape from the gravitational field of the earth has practically ceased and so it is effectively a trapped atmosphere.

Conditions on the sun are quite different: heating from inside (below) is intense and by a variety of processes takes place continuously in depth through the whole of the lower atmosphere, while a supply of material of the same nature as the atmosphere, from further down, is continuously available. Because of the high temperature of this region (the corona) which can reach as much as 10^6 K, the solar atmosphere is, firstly, almost completely ionized and, secondly, not a trapped atmosphere but rather one continuously driven out from the sun and quickly reaching supersonic velocities. It streams past the earth at velocities which range from 300–500 km s^{-1} and goes on much further. Quite how much further is not well known, but almost certainly for perhaps ten times the distance of the earth from the sun. However, eventually its kinetic energy is absorbed in pressing back the interstellar gas which would otherwise occupy the solar system.

D

So we visualize the solar system as a bubble of alien gas within the normal interstellar gas of our particular region in the galaxy. This bubble has a source—the sun, and a sink—the boundary with the interstellar gas—it starts slowly from the sun, attains high radial velocity over a large part of its volume, including the neighbourhood of the earth, and is finally brought up against the restraint of its outer boundary.

Cosmic ray modulation effects arise from details which are really only incidental to the main concept of the solar wind, for there is no reason for the wind, as we have just described it, to have any effect at all upon cosmic rays. The description ' wind ' perhaps conjures up the notion of a dense gas, but this is mistaken, it is extremely tenuous, with perhaps 100 atoms cm^{-3} (to be compared with about 10^5 atoms cm^{-3} in the best attainable vacuum in the laboratory) and almost completely ionized. So the ' mass thickness ' of the solar atmosphere (see p. 17) is very slight indeed and quite unable to absorb any appreciable fraction of cosmic ray primaries coming to it.*

The importance of the solar wind for the flow of cosmic ray particles arises because its ionized gas carries magnetic fields out from the sun and because these are not uniform. The presence of a magnetic field in the wind provides a mechanism which modifies the trajectories of individual particles, but it is the fact that there are irregularities in the field which makes it possible for the trajectories of particles to experience something of the nature of scattering. For this reason, these irregularities are described as ' scattering centres ', and all discussions on solar modulation of galactic cosmic ray particles are based on the strength and distribution of scattering centres.

The idea of scattering centres for solar modulation is an important one, and, moreover, it is a concept that we shall later carry over into interstellar space, so we now discuss it in some detail.

3.3. Scattering in the solar wind

That a magnetic field is non-uniform is not of itself a sufficient reason for charged particles passing through the field to be scattered. For example, the electrons and protons in the radiation belts of the earth† perform a very complicated motion in the geomagnetic field. They spiral about a flux-line; the pitch angle of the spiral increases as the field strength increases from the equator towards the poles, and individual particles are reflected at that point in the field for which the particular pitch angle of each becomes $\pi/2$. Finally, because the lines of force about which they are spiralling are themselves curved,

---

* It is worth making an estimate of the mass thickness of the bubble of solar wind gas. Assume that this is hydrogen and that the diameter of the bubble is ten times the diameter of the earth's orbit.

† See Appendix B.

the whole system rotates about the axis of the earth. The whole motion is a steady and predictable one, with nothing in the nature of scattering taking place.

Now, if, to take an extreme case, the line of force around which a particle was spiralling at some point quite suddenly, in a distance small compared with the spiralling radius, turned through a right angle, then the attachment of the particle to that particular line would almost certainly break down, and where it went next would depend on matters of detail—the exact point on the spiral at which the effect of discontinuity became effective, and size of the discontinuity, the field-strengths in the regions of discontinuity, and the nature of the fields beyond it. Such a discontinuity would constitute a scattering centre.

What has been described may be a limiting case, but evidently less drastic field changes would have to be considered in detail to determine whether or not they were effective scattering centres; whether they scattered at all, and, if so, over what range of angles. The way in which we defined a ' sudden change of field ' above explains how a scattering centre for particles of a certain kinetic energy need not be for those of a different energy, and fig. 11 is a two-dimensional attempt to illustrate this behaviour.

The discontinuities in the fields carried by the solar wind are obviously not the sort of thing for which an exact description is possible, even at a particular instant of time, and so treatments of them have always to be in terms of very much simplified models. For example, one can assign a random distribution of centres, each of which scatters completely (that is to say isotropically about itself) all particles of energy within some range E_1–E_2 entering a certain volume V_0 around the centre, and has no effect at all upon less energetic or more energetic particles, or on any which do not pass through that particular volume. The properties of the medium arising from this scattering is then a function of the distribution of the parameters E_1, E_2 and V_0, and are described in terms of diffusion coefficients. The scattering in any particular part of the interplanetary medium is then described by a diffusion coefficient which is a function of particle energy (strictly of particle magnetic rigidity*). Finally, it has to be remembered that the chosen volume, V_0, is moving along with the solar wind, since it is identified by the magnetic field configuration which the wind carries.

Now let us consider two of the main modulation effects, the diurnal variation and the eleven-year variation.

The effect of the diffusing properties of the solar wind which we have been discussing is more important for low-energy particles than

* See Appendix A.

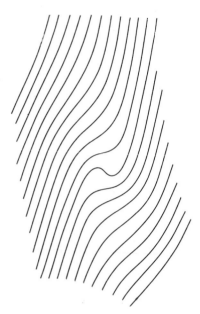

Fig. 11. Two-dimensional diagram of a scattering centre. For simplicity assume that there is no difference of field-line configuration in the direction of the third axis. On the scale of the diagram it is easy to accept that a particle spiralling along any flux-line with a radius of less than 1 mm or of more than 100 mm is unlikely to be scattered. The behaviour of a particle with radius of motion 3–10 mm is less easy to predict. If it is spiralling along the most strongly disturbed line, the reader will probably be able to satisfy himself that the direction of motion of the particle could be changed so that it would return in the reverse direction either along the guiding line to which it was first attached or else along a neighbouring line.

for those of high energy, since the former are significantly deflected by relatively small field variations operating over quite small volumes. The effects are therefore expected to be most noticeable for low energy particles and to become negligible for very high energy ones. Below a certain energy, we can regard the particles as completely diffused in the medium of the solar wind. Individual particles have not changed in energy; what is different from the situation outside the solar system is that the frame of reference in which the total flux of particles of a particular energy is isotropic is now different. It is moving with the gas of the wind, that is to say, outward from the sun.

As a result, in the frame of reference of the sun itself, these particles are moving more readily away from the sun than towards it, and particles entering the solar system from outside are hindered from moving freely towards the centre (e.g. the sun). There is therefore a gradient of particle intensity throughout the space controlled by the solar wind, with the intensity of low energy particles falling off from the external value as they move inward towards the sun. The intensity measured, for example, on satellites near the earth is not as great as it would be were we able to send detectors right outside the solar system. Also, however, at the earth the flux of low energy particles is not isotropic in simple solar co-ordinates—it is isotropic relative to the gas of the near-by solar wind—and as the earth spins on its axis, any detector fixed on the earth scans once each solar day over directions of variable intensity. This is the solar diurnal variation.

The amount by which the intensity near the earth is reduced below that outside the solar system depends on the number and strength of scattering centres in the solar wind. Now these centres have their origin right at the outset when their particular volume of solar atmosphere moves away from the lower layers in which it has been subject to heating processes and to accelerations. The disturbances produced in this way, and then moving outward through the solar system, vary with time, and in particular become stronger and more numerous when the sun is ' active ' than when it exhibits only low activity. It has been known for a very long time that the activity of the sun, as measured by the presence of sunspots, shows a cycle of a period of about eleven years; and the eleven-year variation of cosmic ray intensity, in which the intensity is at its lowest during the period of high activity, when the number and strength of scattering centres leaving the sun is at its greatest, and recovers at the time of sunspot minimum, is only one more physical feature related to a cycle of activity which was already well known.

These two fundamental modulation features are expected to be most marked for low energy primaries, and so a description of the apparatus which has become standard for measurements at the lowest convenient energy at ground level will not be out of place. We can take measurements with such equipment as a basis, and then go on to consider whether the reduction of these effects at higher primary energies can be established. Such measurements taken altogether lead to qualitative information about the scale of irregularities in the magnetic fields of the solar wind.

3.4. The neutron monitor

At first sight there would be no question at a sea-level observatory of making any measurements on primaries of energy lower than about 2 GeV since (p. 18) we have seen that a proton would lose that amount of energy just by ionization in passing through the atmosphere even if it suffered no other interactions at all. Heavier nuclei would lose

43

still more. In fact primaries of even lower energy do produce sea-level effects, and the simplest mechanism by which this can happen is if, in an early interaction, neutron secondaries are produced which either themselves penetrate right down through the atmosphere, or else in one or two further interactions yield yet other neutrons which do so. What is necessary is that this individual path within the nucleon cascade should have been travelled almost all the way as a neutron, which does not lose energy by ionization since it is not charged. The sooner the primary proton interacts to yield a neutron the better, and while the last particle of all may again be a proton, it will presumably have only been formed in a very late interaction. For the most sensitive modulation measurements, therefore, we use detectors sensitive to nucleons only, since the much more numerous

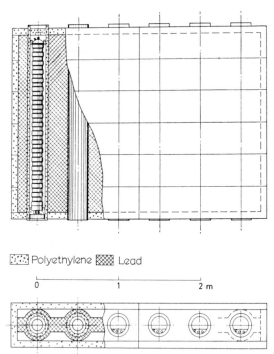

Fig. 12. Plan and end elevation of a six-counter unit of a NM64 nucleon monitor. Incident nuclear particles interact with the lead target yielding evaporation neutrons which are moderated in the inner poly-ethylene sheath and captured in the gas of the counters. A full monitor consists of three units: by continuous intercomparison of counting rates in the three sections any defect of operation is identified with a minimum of delay, and since it is extremely unlikely that two sections will show any defect at the same time, good data are to be expected from the remaining two sections.

44

muon and electron secondaries (about 99% of all particles) which reach the ground have all certainly arisen from primary particles of higher energy. It is true that, particle for particle, high energy primaries also produce greater numbers of secondary nucleons at sea-level, but this effect is more than offset by the much greater intensity of low-energy primaries.

Figure 12 shows, in plan and front elevation, one six-counter section of the current standard detector for modulation phenomena, the ' NM64 ' neutron monitor, while fig. 13 (a) shows two sections of an actual monitor and fig. 13 (b) the control and recording system for it. The name arises from the detailed part that neutrons released within the monitor play in its operation and does not indicate that it is sensitive only to neutrons falling on it (NM64 refers also to the year in which this instrument was adopted as standard equipment). Strictly it is a ' nucleon detector ' and it is equally sensitive to incident protons and neutrons, and indeed to pions, although because of the requirements of passing through the atmosphere with minimum ionization loss, many more neutrons than protons or pions arising from low energy primaries reach the monitor.

The apparatus consists basically of a lead target in which incident nucleons undergo nuclear interactions with the very large heavy nuclei of lead. A major product of these interactions is a number of ' evaporation ' neutrons, released very much as evaporating water molecules would escape from a small strongly heated water drop. It is these neutrons that must be detected, but at the evaporation energy (~ 1 MeV) really efficient detectors are not available, so within the lead target there is a moderator, here polyethylene, in which these neutrons are slowed down by elastic collisions with hydrogen nuclei. Within this moderator, again, is a specialized counter in which low energy (moderated) neutrons can interact and be distinguished. These counters have as the main filling the gas $^{10}BF_3$, incorporating the boron isotope of mass ten, which has a very high probability for slow neutron capture according to the reaction:

$$^{10}B(n, \alpha)^7Li + 2 \text{ MeV} \quad \text{(as kinetic energy of the products).}$$

The recoiling products of this reaction produce very heavy ionization in the gas of the counter and, when this is used as a proportional counter,* can be detected separately from all other ionization taking place within the counter by the very large pulse which each neutron capture produces.

*A proportional counter is structurally like any Geiger counter, but is operated under different and stringent conditions so that the signal produced on each occasion when it is discharged is proportional to the initial amount of ionization produced in the gas of the counter.

(a)

(b)

Fig. 13. (a) Two six-counter sections of an NM64 monitor (the corner of the third section is just seen, bottom left). The general scale of the instrument can be judged from its surroundings. (b) Control and recording equipment of an NM64 monitor. The automatic barometer is mounted on the white plinth back right of centre. The recording system is at the extreme right. An automatic typewriter records every five minutes, and identical data are punched on to computer tape.

From fig. 12 it will be seen that the lead target is itself surrounded by a further moderating layer. This acts partly as a reflector sending back evaporation neutrons which began to move outwards rather than inwards from the nuclear interactions, but it also acts as a shield to prevent changes in the immediate surroundings of the monitor altering the conditions inside it. For example, it minimizes the effect on the counting rate of people working near the monitor during the day and not being there at night, and of snow, and rain upon the nature of its surroundings.

We have explained the advantages which there are in using a detector sensitive to the nucleonic component for studies of variation of intensity which most affect low energy primaries, but there is, at first sight, one very serious practical difficulty. Nucleonic particles make many collisions and so the intensity observed at a neutron monitor depends very much on the mass of air above the apparatus: the greater the mass thickness the more collisions will have taken place, and the more the nucleonic component will have been attenuated. The effect is a large one, for each millimetre of mercury change in barometric pressure the nucleon intensity varies by about 1% and from the highest to the lowest barometric pressure which one may expect over a few months, this effect alone can change the observed intensity by a factor of almost two! Fortunately, providing one knows the barometric pressure accurately, as a measure of mass thickness of air, a quite exact correction can be made. The present NM64 monitors, as normally operated, incorporate an automatic recording barometer stable to about 0·1 mm of mercury; the pressure is printed out with the counts of the three channels of the monitor every five minutes and is also included in the computer tape punched for each of these intervals. In what follows the data discussed will always be that corrected for barometric variations but it is useful first to get some sense of the size of this correction compared with the details of intensity which are of actual interest. Fig. 14 shows uncorrected and corrected data at Leeds for about three months in 1966, and in addition shows the (corrected) data for the important pioneer Canadian monitor at Deep River. The general similarity of the corrected Leeds and Deep River records is obvious: closer examination does show differences; these are important insofar as they record the extent to which the movement of particles near the earth is not isotropic.

Figures 15 and 16 illustrate data which bring out some important features about the solar diurnal variations of intensity. They refer to two short periods of 'quiet' conditions in May and December of 1969, as these were recorded by the Leeds NM64 monitor.

Solar weather, like our own, is never 'quiet' to the extent of absolute stagnation, and these examples quite obviously include

47

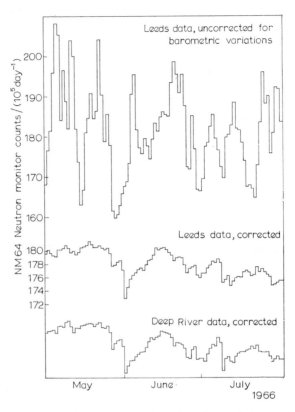

Fig. 14. Correction of neutron monitor data for variations of barometric pressure. The barometric pressure is effective as the measure of the mass thickness of air over the apparatus, and the magnitude of the correction is determined from extended correlation of pressure with monitor counting rate. Leeds data are compared here with data from the Canadian monitor at Deep River; the uncorrected data for Deep River were not available nor the numerical value of the corrected rates. The broad similarity of the data is clear; significant differences can also be noted.

variations other than a simple diurnal effect. But they are very undisturbed compared with the conditions which we will illustrate in a following section, and in figs. 18 and 19, where it is not surprising to find that the diurnal effect we are now examining is quite swamped by violent variations.

In fig. 15, the quiet periods 4th–9th May and 8th–15th December, 1969 are plotted for (barometric) corrected data, each step of the plot being an average of three hours counting. Two features as well as the diurnal variation are noticeable. Firstly, there is a long-term drift

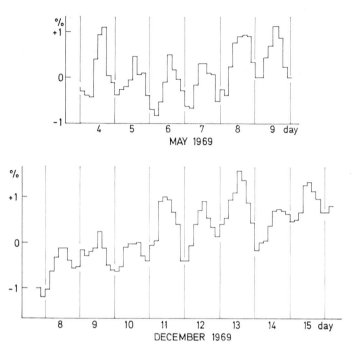

Fig. 15. Corrected NM64 data for 'quiet' conditions in May and December 1969 showing variations of the order 1%. A diurnal variation is superposed on a long-term drift, which differs in the two examples, while in turn the diurnal variation is distorted by smaller effects which do not repeat day by day.

which for the May plot appears as a sort of trough and for the December one is a steady increase of intensity. Secondly, there are certainly other short-term variations affecting the record so that the 'shape' of the recorded variation varies from day to day. Regarding this shorter variation, it may be worthwhile to look ahead to fig. 20 (b) in which a change larger than any whole diurnal amplitudes that we are now studying is shown taking place in a few minutes! To bring out more clearly a probable periodic effect against other possible sorts of variation, we can add together a series of cycles of the known period. Then other effects should relatively be smoothed out and the periodic effect should show up more clearly. This has been done in fig. 16. Here we have used hourly averages rather than three-hourly ones, and made six-day and eight-day 'running averages' for the May and December periods respectively. Thus we have added for the May measurements the intensity for 0–1 hours, 1–2 hours and so on for each of six successive days, and as was to be

49

expected, the diurnal period shows up more strongly and we are able to make meaningful estimates of its amplitude and time of maximum. Averages over much longer periods would be needed to determine its 'shape', but the general form of the curve is that of a single wave in one day, which means that the 'first harmonic' is probably the important one.

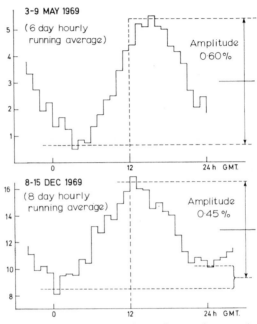

Fig. 16. The data of fig. 15 averaged for each set using a period of one solar day. This procedure enhances the diurnal variation relative to small non-periodic disturbances, and allows meaningful estimates of amplitude and phase. For the December data this averaging process does not cut out the fairly steady drift of intensity which was apparent over the whole of this period. As shown in the diagram, the intensity maximum is compared, for the purpose of making an estimate of amplitude, with the average of the intensities 12 h earlier and 12 h later in this averaged plot.

The phases of both curves agree in putting the maximum a little later than local solar noon, and if we were to collect comparable records from stations, for example, in America, they would also show this feature relative to their own time and local noon. The phases of maximum are not exactly the same for both curves, and indeed if we examine a great many like them we would find the maximum drifting about a bit, but in our examples we have only averaged over quite short periods, and rather untypical and disturbed variations on

50

9th December will have tended to shift the maximum of the second average noticeably towards an earlier time.

In taking an average over periods of six or eight days respectively we would not expect to distinguish between a solar effect and a sidereal effect: both would have added up closely in phase as has happened in fig. 16. However, if the phase of maximum was more or less steady in sidereal time rather than in solar time, between May and December sidereal time and solar time have changed relative to each other by about 12 hours, and the two curves should be out of phase with each other rather than in phase. These two curves alone thus make it likely that this is a solar diurnal effect rather than a sidereal variation. By examining more such periods this conclusion would soon be firmly established.

The amplitude of the variations seems to vary to some extent, but the data shown here is not enough to establish whether there is a real difference or whether the apparent difference is not the result of the different degrees to which disturbing effects have been cancelled out by taking a 'running average'. When the effect we are examining is superposed upon a longer term variation as it was in December, fig. 16 shows quite clearly that the procedure we have adopted does not eliminate it, and the diagram also shows how a rough allowance for it has been made.

Before we summarize what is to be derived from our study of these two 'quiet' periods, a word of caution is necessary. Because of the heavy atmospheric absorption of nucleons, which leads to the large barometric effect already described, the directions of approach of particles recorded by a sea-level neutron monitor are concentrated near the zenith, and the whole flux of particles can be imagined as coming down in the vertical direction and apparently defining clearly the direction in space from which they have come. However the geomagnetic field cannot be ignored, least of all for the decidedly low energy primaries which are studied using neutron monitors. The apparent direction—that at the top of the atmosphere and within it— can be quite different from the asymptotic direction of a particle trajectory, the direction in which the particle was moving before it began to be deflected by the earth's magnetic field. Moreover, the asymptotic directions of particles leading to neutron monitor measurements can be spread over a very wide range of directions according to the energy of each primary particle involved (fig. 17). So the relationship of direction, and even to some extent the amplitude, of the diurnal effect with the proposed mechanism to account for this effect is in reality a decidedly complicated one involving a great deal of detailed computation. The diurnal variation observed at a particular place using a standard neutron monitor is an effect related to a range of primaries coming from different asymptotic directions according

51

to their energy, but it is possible to regard it as mainly deriving from primary energies around a certain average value and coming from some general direction. It is important, therefore, to know how the diurnal effect varies with the direction in solar space which is scanned and particularly with the average energy of primary which contributes to the measurement.

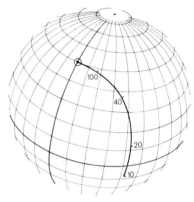

Fig. 17. This diagram shows a view of the northern hemisphere marked off in 10° intervals of (geomagnetic) latitude and longitude. The geomagnetic equator and meridian through the impact point are marked with thicker lines, and the impact point is, for this illustration, in 50° north latitude. The very heavy line springing from the impact point links together differing primary energies (GeV), and shows the *directions*, far from the earth, in which particles have to be moving if they are to reach the earth vertically at the impact point. Primaries of extreme energy will come without deflection but there is significant deflection even at 100 GeV (10^{11} eV) and so great are the deflections at low energies that no attempt has been made to follow them below 10 GeV. It is important to understand that the various points on this locus describe *directions* far from the earth and do *not* indicate that these particles were 'aimed' towards the actual points on the earth with these angular co-ordinates.

Both of these variables can be altered in two ways: by making observations over a range of latitudes, and by altering the energy-sensitivity of the actual detector. However, in each approach there are serious complications, since *both* average energy *and* asymptotic direction of approach vary together in a way which is not under control, and, moreover, these averages conceal quite wide ranges both of energy and direction. The most clearly known feature at present is the comparison of the diurnal effect at a single place measured on the one hand in a neutron monitor and on the other in a counter telescope shielded so that it receives only the muon component. In middle latitudes, and ignoring the difficulties about spread of energy and of direction we have just been discussing, the main sensitivity of the neutron monitor is for primaries of perhaps 1–3 GeV and of the muon telescope for about 10 GeV.

The diurnal amplitude in the muon detector is found to be about half that in the neutron monitor: if the muon detector is operated underground, its sensitivity can be shifted to still higher primary energies, since the detected muons must have been all of high energy at ground level if they are to be able still to penetrate considerable depths of rock, and in this way the diurnal effect is detectable, although now much smaller, even when the average primary energy approaches 100 GeV.

When all this experimental material is brought together, it emerges that the ' true ' direction of maximum intensity outside the influence of the geomagnetic field is rather later than our neutron monitor examples indicated. It lies at about 1800 h local solar time, and this direction is in fact that at which the solar wind overtakes the earth in its orbit. It comes in this direction because the wind ' corotates ' with the sun: from a point on the sun the wind would appear to go out radially, but from the earth, and because the sun is itself rotating about its axis relative to the earth–sun line, the direction of the wind must appear as a spiral, and this is the direction (along the spiral) which we are able to determine.

3.5. *The Forbush effect*

The field distortions, which yield scattering centres and lead to the diffusive nature of the motion of low energy primaries through the solar wind, can pass near to and even envelop the earth itself, and we can regard a great deal of the variability of the day-to-day effect as coming from the special influence of such near events. However, when an exceptionally large solar disturbance takes place (one of the so-called ' solar flares ') and a major outgoing distortion of the wind sweeps over the earth, modulation on quite a different scale takes place. Then there is a sudden decrease of intensity which may be by as much as 10% or more, although smaller decreases from the un-observable upwards are constantly occurring, this is the Forbush effect, named after the American, Scott Forbush, who pioneered modulation studies and who first identified this particular feature.

A striking example of the Forbush effect is shown in figs. 18, 19 and 20 from observations at Leeds in October and November of 1968. Fig. 18 shows the intensity averaged over whole days for a period of about a month; it shows an effect lasting for more than a week, and quite out of scale with the quiet day-to-day variation which preceded it. Fig. 19 shows an hourly plot of the onset of this particular event: it shows that the time of onset can certainly be fixed to within an hour, and that the intensity went on falling at a rate of about half a per cent

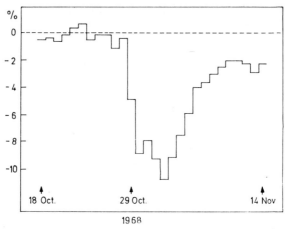

Fig. 18. The Forbush event on 29th October 1968 (NM64 data). The graph shows the daily average intensity over about a month related to the long-term average at that time as zero.

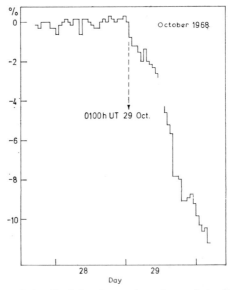

Fig. 19. Onset and detail of the decreasing phase of the Forbush event on 29th October 1968 as shown in one-hourly average intensities. The time of onset to within one hour is clear, as is the fact that this is abrupt to this scale of timing.

per hour for almost a whole day. Finally, fig. 20 (*a*) shows the most rapid period of decrease in this event plotted in 5 minute intervals. Here we see that even on this scale there are obvious irregularities, with a maximum rate of decrease extending over perhaps 20 minutes of about 10% h^{-1}. With a still larger monitor and a much greater counting rate, even finer details would probably have been identified, since here we are reaching the statistical limit of resolution of the present monitors.

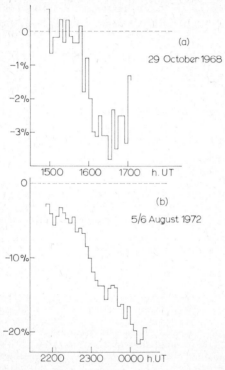

Fig. 20. (*a*) The period of most rapid decrease in the event of 29th October 1968 plotted in 5-minute averages. Here the statistical fluctuation of the counting rate is becoming significant, and no finer time-division would have brought out any further real detail. (*b*) The period of most rapid decrease of a more recent Forbush event (August 1972) which led to an exceptionally large total reduction of measured intensity. This, as in (*a*), is a plot of 5-minute averages.

Fig. 20 (*b*) shows a similar 5 minute interval plot for one of the largest recorded examples of the Forbush effect, which has occurred more recently (August 1972). Here the measured rate decreases by over 5% in 15 minutes, and so by the average amplitude of the whole diurnal effect in perhaps three minutes, more than twice the maximum rate measured in the event which we have examined in detail.

Changes during the recovery phase are much slower, the fall of intensity which takes place in about a day takes a week or longer to recover, and indeed on some occasions the recovery seems never to be complete. If the irregularities which lead to the diurnal effect are ripples in the solar wind, the disturbance leading to this Forbush event is something like a tidal wave. It swept particles out of the neighbourhood of the earth so decisively that this long time elapses before they have even approximately established themselves there again.

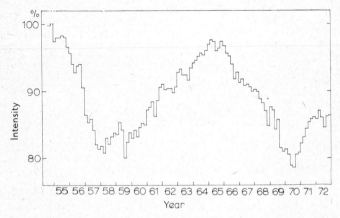

Fig. 21. Graph showing the eleven-year periodicity of cosmic ray intensity as recorded at Leeds. About 18 years are covered here and these include the times of maximum solar activity in 1959 and 1970. The data collected here, showing the minimum intensity in 1970, the climax of a relatively mild cycle, actually lower than that of 1959 in a very active cycle are notable. Taken in conjunction with the fact that the intensity did not completely recover after the 1954–1965 cycle, this suggests that some disturbances may have remained from this cycle into that which followed.

In fig. 21 we show solar modulation on yet another scale, and this figure covers a period of about 18 years, with two-monthly averages plotted over this period. In looking at this plot, it has to be remembered that we have here a smoothed-out version to which all the detail we have just considered, even to that which happens in times shorter than an hour, has contributed. While fig. 21 brings out the 11-year variation of intensity, it does so only in terms of a little more than one cycle, for unfortunately comparable measurements do not go back further. But already it is clear that one solar cycle is, as far as cosmic ray intensity is concerned, not going to be exactly the same as another, and it will obviously be a long time before we can add much to this statement. Cosmic ray intensity varies more or less inversely with the number of sunspots, and the reason for this will have become apparent in the paragraphs above. The sunspots are, if you like, a symptom of the sort of disturbed conditions on the surface of the sun which leads to the emission of stronger and more numerous scattering centres into the interplanetary medium, and so it is when the activity of the sun is high as judged by the presence of sunspots that the sweeping effect of the rough wind, tending to keep low energy particles out of the inner solar system, is at its strongest.

Surveying the whole of modulation phenomena one sees that they are indeed a sort of description of weather in the solar wind, ranging from broad periods of storm to relative periods of calm, from intense individual storms to the general uncertainties even of relatively fine weather. One has to notice that what we observe on the earth is not, as regards the weather, a point measurement, although we of course are making a point measurement of cosmic ray intensity. This intensity has been affected by the conditions in a very large volume indeed surrounding the earth, and so in a way it represents an average of weather conditions. In this, it must be contrasted with measurements that can be made on satellites and space probes, because here one is making instantaneous measurements (for example, of the interplanetary magnetic field) at a single point. But it is a point that one has not got under one's control with regard to a particular phase of the solar weather: we cannot stop the vehicle in its course to keep it at a place when particularly interesting conditions are developing. So these two methods of measurements are complementary, cosmic ray measurements on the earth yielding information about average conditions in the weather and satellite measurements yielding detailed point-by-point measurements, and allowing us to see actual examples of the sudden changes in the direction and intensity of magnetic field at certain points and times which we have identified as the simplest form of scattering centre.

CHAPTER 4

the charge and energy spectra of primary cosmic rays

4.1. *General features of spectra*

THE data which we shall now discuss are perhaps the most important features of the subject as they relate to general cosmology. They are what we can actually measure about the relative abundance of the various kinds of particle, and the energy spectrum (the differing proportions of particles of a given species having particular energies) for each of these. At the lower energies, up to perhaps 10^{12} eV, where adequate measurements are possible at balloon altitudes and beyond, we now have a great deal of information about the energy spectra of individual kinds of particles; protons, alpha-particles and heavier nuclei, also of electrons. As the primary energy increases, our ability to distinguish the effects of different kinds of particles quickly becomes much less effective, but we can still refer to the energy spectrum of cosmic ray primaries as a whole, for most 'shower' measurements, for example, are roughly indicative of the total energy of the primary particle, and the degree to which these can be refined to give precise estimates is perhaps surprising.

In discussing energy spectra there are possible points of confusion, and so some definitions are first necessary. The *integral* energy spectrum of a particular kind of particle (at a given place) describes the relative numbers of particles which have greater than a chosen energy, and we can write this as $I\,(>E)$. It may either be a measure of the directional intensity, being defined in a particular direction and within a particular solid angle, or of the total omnidirectional intensity from all directions. We may take as a unit of intensity the number of particles crossing unit area in unit time within unit solid angle, and so we have the intensity:

$$I\,(>E), \text{ measured in } \mathrm{m^{-2}\,s^{-1}\,sr^{-1}}$$

where the steradian (sr) is unit solid angle.

Whereas on satellites it is just possible to imagine a detector sensitive omnidirectionally, this plainly is not so on the earth where at best one hemisphere is open, and there are very few situations where a specific directional intensity is not clearly preferable. As a rule, therefore, and unless it is qualified, $I\,(>E)$ refers to a directional intensity and, unless otherwise stated, to the vertical intensity. Non-vertical intensities may be indicated as $I_\theta\,(>E)$ where θ is the zenith angle of the direction of observations. Occasionally the azimuth angle has also to be indicated.

There is no obvious analytical form to be related to the energy spectra and we do not understand enough about them to expect that there should be, but it turns out on inspection that over wide ranges of energies the spectra approximate to a power law function:

$$I\,(>E) = KE^{-\gamma}.$$

Here the constant K describes the level of intensity of this particular group of particles and γ the ' shape ' or ' slope ' of the spectrum; for since cosmic ray spectra are often displayed over ranges of energy of many powers of ten, they are usually plotted using logarithmic scales, both for I and for E. Then $-\gamma$ is the ' slope ' of the curve at any point, and it is very often referred to as the ' slope of the spectrum '.

This proves to be a most useful way of describing spectra, and with it the finer details of a spectrum can be brought in by allowing γ to be a slowly changing variable rather than a constant. If we follow the spectrum of cosmic ray primaries from a low energy, say 10^9 eV, up to a very high one, perhaps 10^{19} eV it will be described by values of γ appropriate in different parts of the range.

What we have described is an integral spectrum, defined by the intensity of particles with energy greater than a particular value, and it is what comes out of the most straightforward treatment of actual observations. However, for comparison with theoretical predictions a differential spectrum is often more useful. In the simplest situation, where the integral slope, γ, is a constant for all E:

$$\mathrm{d}I_i = -I_d\mathrm{d}E; \quad I_d = -\frac{\mathrm{d}}{\mathrm{d}E}KE^{-\gamma} = K'E^{-(\gamma+1)},$$

where I_d and I_i are respectively the differential and integral spectra. That is to say, the differential slope is greater by one than the integral slope. Its units include a range of energy ($\mathrm{d}E$) for which the constant K' must be defined, and these will accordingly be of the form $\mathrm{m}^{-2}\,\mathrm{s}^{-1}\,\mathrm{sr}^{-1}\,(\mathrm{MeV})^{-1}$, where some such convenient energy step, $\mathrm{d}E$, is chosen.

Finally, when we want to consider the spectra of different primary particles, and particularly how they move in magnetic fields and how the product particles move after collisions, it is useful to introduce the idea of the ' energy per nucleon ', and then the differential spectra will refer to kinetic energies measured in, e.g., GeV (nucleon) $^{-1}$. In fig. 22, the differential spectra, per nucleon, of some nuclei in the range of energies where it is possible to carry out balloon or satellite experiments which distinguish between various nuclei of the primary radiation, are plotted.

59

We use these at the moment only to lead into a discussion of the nuclei which make up the primary radiation. Because it appears that over the energy range covered in this graph the relative abundance of the various nuclei seems to change very little, for these energies it is possible to refer to a ‘ charge spectrum ’ or ‘ nuclear composition ’ of primary particles without having to relate this to a particular energy. The ‘ nuclear composition ’ is only that for particles in this particular

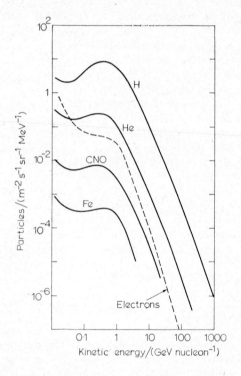

Fig. 22. Energy spectra per nucleon of primaries over the range for which these particles can be directly detected in balloon or satellite-carried equipment. The curve marked ‘ CNO ’ refers to almost equal amounts of carbon and oxygen with a small admixture of nitrogen; that marked ‘ Fe ’ does not distinguish between iron and its immediate neighbours. The lower ends of these spectra (energy per nucleon less than 10 GeV) are strongly modified by solar modulation: the amount of depression of intensity here varies through the solar cycle, and at the lowest energies some particles originating at the sun are also present. Outside the influence of the solar system the spectrum above 10 GeV nucleon^{-1} would probably continue smoothly to lower energies (see fig. 25). The broken line refers to the electron spectrum, and for this curve the units will be GeV and $m^{-2} s^{-1} sr^{-1} MeV^{-1}$.

60

energy range, and must not be presumed to apply without question to primary particles of much greater energy. However the energies to which it does apply are important ones, for it is in this range that most of the energy which exists in the form of cosmic rays seems to be found.

These primary particles, for which the various elements are distinguishable, have a general interest too: they provide us with the only sample of matter of measurable composition available from outside the solar system. This sample is, we of course believe, selective from particular environments, but its unique information is nonetheless important for that.

4.2. *Nuclear composition*

What is known about ' nuclear composition ' in the sense which we have just established is shown in fig. 23. In using this diagram it is important to notice that there is a gap in the scale for both co-ordinates and a change of scale for one of them. It brings together conclusions

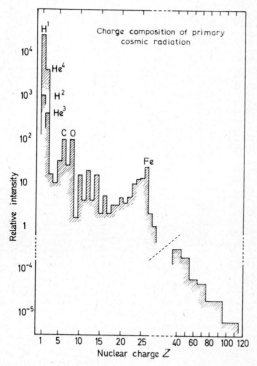

Fig. 23. Relative charge composition of primary particles near 10 GeV/nucleon. For discussion, see text, but note particularly the reduction of intensity, by perhaps three orders of magnitude, beyond the iron nucleus.

61

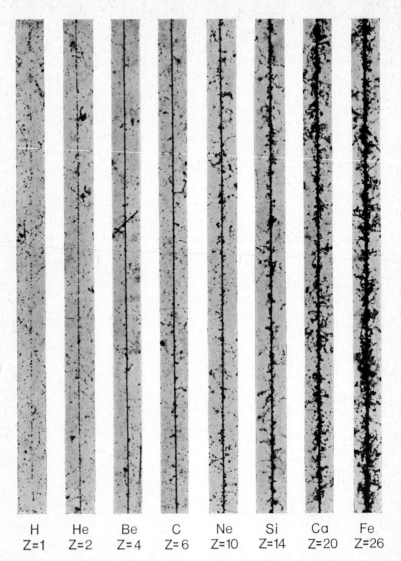

H	He	Be	C	Ne	Si	Ca	Fe
Z=1	Z=2	Z=4	Z=6	Z=10	Z=14	Z=20	Z=26

Fig. 24. Tracks in nuclear emulsion left by relativistic nuclei, $Z=1$ to $Z=26$. From very high balloon flights and from satellite exposures these will almost certainly be primary particles. The photographs each show slightly less than $\frac{1}{2}$ mm of track length. The loss of energy by ionization which leads to the activation of grains of silver bromide in the emulsion is expected to be $\propto Z^2$. While this cannot be established by scrutiny of these tracks, the blackening is certainly increasing more rapidly than $\propto Z$.

from the application of a great many techniques—photographic emulsion (see fig. 24), ionization damage in various materials (which is an approach which promises to be most important for very heavy primaries) and a variety of electronic devices which measure the two necessary quantities, E and dE/dx in a more or less direct way. This is a part of the subject which is developing very rapidly, and what is known will have quite certainly grown dramatically before what appears here is published.

The break in fig. 23 immediately after iron is one of confidence as well as of scale, for the very rare, very heavy nuclei beyond this interruption are identified with much less certainty than are those in the lighter group. It seems, however, that these include, in a small number of instances, nuclei of charge comparable to that of the heaviest known on earth. A few have even been interpreted as ' transuranic ' (heavier than any known nucleus of earth matter), but the evidence for this identification is not altogether convincing.

The lighter elements, however, present a much clearer picture, in which the most prominent feature is the predominance of the stable light elements, carbon and oxygen and the outstanding importance of iron. This particular feature accords with the well-known position of the iron nucleus as probably the most stable configuration of nuclear matter, while the pattern set by carbon and oxygen is followed, less prominently, by a succession of other ' even-Z ' nuclei, which consistently are several times more abundant than their ' odd-Z ' neighbours.

The general picture is one of nuclei in the sort of abundance in which they must be formed in catastrophic cosmological events such as, for example, supernova explosions, and with a composition quite different from that of the general interstellar gas, which is almost entirely hydrogen and helium.

For the lightest elements, up to boron, a further feature must be noticed. The isotopes of hydrogen and helium ^{2}H and ^{3}He, which are rare in these gases on the earth, are relatively abundant, but still more striking is the presence in far from negligible quantity of the elements lithium, beryllium and boron. Striking, because these three elements are all extremely unstable under the conditions of formation necessary for the heavier elements: indeed even under the mild conditions on the sun they can have only transient existence. There is no serious doubt that these various unexpectedly abundant light nuclei were not present at the original formation and acceleration of the main bulk of the primary particles, but are fragments knocked off larger nuclei during their passage through the interstellar gas.

The composition of the primary particles when they reach the earth accordingly carries information *both* about the conditions under which particles of cosmic ray energies are formed and *also* about their subsequent history, particularly about the amount of interstellar matter through which they appear to have passed in the process.

4.3. *Energy spectra*

What we know about the energy spectrum of the primary particles falls into two distinct regions and it is important to emphasize the implications of this separation, and the way in which the two ranges of energy which they cover are to be related.

Data relating to rather low energies (to beyond 10^{12} eV) have already been given in fig. 22. Here primary particles are sufficiently abundant for high balloon and for satellite work to be carried out which distinguishes directly between the various kinds of nucleus reaching the earth, and the graphs show a set of energy spectra (strictly, plotted as energy per nucleon) which refer to the more abundant sorts of nucleus. All of these are roughly of the same form and at the highest energies are tending towards a slope which is only slowly varying.

At lower energies, $\sim 10^8$ eV per nucleon and below, these spectra show a more complex shape and, further, here there are considerable changes during the eleven-year solar cycle. A great deal of this shape is determined by the modulating function of the solar gas and measurements at the earth, or anywhere near it, give little information about the slope of the various spectra for these energies outside the sphere of influence of the sun. The most likely forms for protons and for electrons are illustrated in fig. 25, which uses modulation functions chosen to fit data for a range of nuclei. What is actually observed at the very lowest energies (10^6–10^8 eV) probably includes a large proportion of particles of direct solar origin.

We may summarise the detailed variations in this low energy region as indicating that it is not until energies of about 10^{10} eV are reached that what is measured near the earth represents substantially the interstellar radiation, and that the likely non-modulated low energy spectrum outside the solar region is an extension of what is measured near 10^{10} eV but with a slowly diminishing spectral index. The slopes of the energy per nucleon spectra, as far as these are known, suggest that, again near 10^{10} eV/nucleon, the slopes of the spectra for the various nuclei are similar.

When we try to extend these spectra to higher energies, we very quickly come to a point where individual primary particles become so rare that there is no sense in which these can be used to describe a series of spectra. The very highest energies at which primary particles have been observed at all are in the region of 10^{14}–10^{15} eV and then only in very small numbers. From the totality of existing observations the proton spectrum can be extended to about 10^{12} eV, and what little is seen of heavier nuclei would not contradict the view that the 'nuclear composition', well defined at 10^9–10^{10} eV, still applies up to

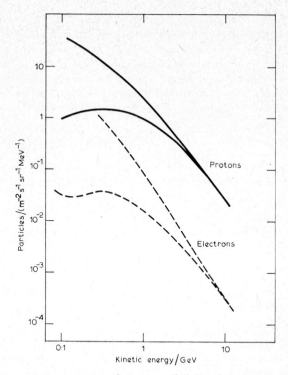

Fig. 25. Low energy spectra of protons (full lines) and electrons (broken lines), below kinetic energy 10 GeV. For each, the lower curve is typical of spectra measured close to the earth in satellite experiments, and would be subject to variation because of solar cycle modulation. The two upper curves are estimates of the situation corrected for the modulation effect of the solar system. They represent the best informed ' guess ' as to the spectra well outside the solar system but none-the-less in our immediate vicinity.

this approximate limit. Over the (proton) energy range 10^{10}–10^{12} eV, the differential slope of the spectrum is changing only slowly, and lies between $-2\cdot6$ and about $-2\cdot7$.

4.4. *Energy spectra derived from measurements on the secondary component*

Beyond the observations described in the last paragraph, our knowledge about the primary particles is indirect: it is derived from studies of the secondary radiation and these do not give any straightforward indication of the nature of the primary particle. Accordingly, what is derived in this region of higher energy is, in the first place, the total energy of a primary particle, and the spectrum deduced is a total-energy spectrum. Fig. 26 shows the integral total-energy spectrum as we now think it extends upwards from near 10^{12} eV till near 10^{20} eV it becomes meaningless to attempt to describe a spectrum for exactly the same reason that near 10^{12} eV it ceased to be possible to describe spectra of separate species of nuclei. Beyond 10^{20} eV only a very few observations have been made, and it is safe to say that up to now no primary of energy as great as 10^{21} eV has been

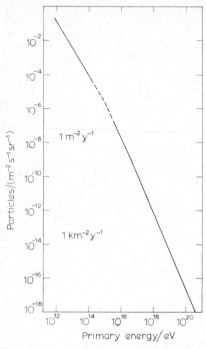

Fig. 26. Primary spectrum above 10^{12} eV; note that this is an integral spectrum, describing the flux of all particles of the given energy and greater, and that it is of *total* energy irrespective of the nature of the primary. Two typical 'practical' rates are indicated. The section of the spectrum shown in broken line is particularly uncertain. Here observations are very difficult and different experiments are not concordant.

recorded. Equally, however, it should be stressed that this provides no evidence that such particles, and others still more energetic, do not occur at rates compatible with what has already been established, but these rates fall right below what could have been expected to yield any observations at all with the experimental dispositions at present available and the time for which they have run.

The problem as to the nature of the primary particles over the very wide range of energy covered in fig. 26 is, of course, of the greatest importance, but one very quickly becomes aware that this importance is at least matched by the extreme difficulty in making any measurements which will provide convincing evidence about it.

The most important feature of the spectrum in fig. 26 is its simplicity. It seems to continue up to about 10^{15} eV with a slope very close indeed to that established directly up to 10^{12} eV, and above about 10^{16} eV it goes for another four decades with an almost constant slope which is, however, about 0·6 steeper than that at the lower energies, say $-3·2$ (differential) as compared with $-2·6$ (differential).

At an earlier stage (p. 33) we have made the rather obvious point that deductions about the primary particles from observations of secondary phenomena are a second best, but here we have a situation in which they are unquestionably necessary, for there is no alternative. Since the spectrum given in fig. 26 is such an important feature in discussions about the cosmological role of cosmic rays, the difficulties of working on secondary data have to be examined in some detail.

There are three main sources of uncertainty in relating secondary phenomena to primary particles:

(i) It is no longer possible to assume the nature of the primary particle: to determine *what it is*, is now one of the (more difficult) problems.

(ii) Over almost all the energy range involved, there are no experimental data whatever as to the result of the early collisions (most of all the *first collision*) of primaries with the nuclei of the atmosphere. Existing data from machine studies have to be exterpolated over many orders of magnitude, but there is not much confidence in this procedure.

(iii) The early collisions (and again most of all the first) occur statistically. If on average, the first collision of a particular species of primary takes place at depth x g cm^{-2} into the atmosphere, it will in fact often take place at $\frac{1}{2}x$ g cm^{-2} or at $2x$ g cm^{-2}, and the first collision at $3x$, $4x$ and even $5x$ g cm^{-2} will not be unknown. It is to be expected that the secondary component, wherever it is observed, will show differences (fluctuations) which arise from the purely statistical differences of behaviour of primary particles in their early penetration of the atmosphere.

4.5. *Model simulations*

These complications would have been daunting had they arisen at a time when computing aids were non-existent or primitive. Now the situation is less difficult, because it has been possible to develop a process of 'model simulations' in which artificial showers are generated by following through a set of imaginary interactions which are made to occur at random within a chosen mean free path, and yielding products, also with a random element, according to the particular postulates about nuclear interactions which are under examination. For a variety of conceivable primary particles it is possible to calculate through, using various assumptions under heading (ii) and as to the mean collision depth, x, under (iii), and to repeat this calculation many times to cover the effect of path fluctuations indicated in (iii) and perhaps also variations of behaviour which may be expected in the collision processes involved in (ii).

Providing that the quantities used to set up the ' model ' are varied in a way which effectively brackets what happens in reality, this simulation approach is extremely powerful. It can in each instance predict a number of observable features of the secondary radiation which can be tested as a group, and which will tend, as material becomes available, to concentrate attention on promising models and to exclude those which regularly lead to predictions apparently inconsistent with what is observed. The weakest link lies certainly in (ii), and concerns the nature of the ' first collision '. One cannot exclude the possibility that at very high energies these collisions are not like anything one has imagined, and so can in no sense be regarded as ' bracketed ' by the alternative models developed.

Another limitation to the simulation approach is the extent to which it consumes computer time, but, in common with many other long computer operations, the nature and application of permissible ' short-cuts ', and concentration of output information on results which are useful for comparisons with important features of the real secondary radiation have been developed, and these give valuable economies.

The upper part of fig. 26, which goes on beyond that region of the spectrum where primary particles are measured directly is the first in which data from secondary particles becomes important, and here what is probably the most reliable derivation comes from measurements of the momentum spectrum of muons at ground level. Some of the most advanced work on the spectrum has been done at Durham, and one model of a series of ' momentum spectrographs ' used for this purpose is shown in fig. 27. Muon trajectories which pass through the air-gap of a powerful electromagnet are selected by five layers of Geiger counters (note the small layer in the actual gap of the magnet) and then four layers of detector-systems which define the trajectory more exactly, once the passage of the muon has activated their operation, are used to obtain the angle through which the muon trajectory is deflected in the magnetic field.* The calculation of

* The high-resolution detector units cannot be shown in any detail in a diagram on the scale of fig. 27. Similar units are shown more clearly and explained in fig. 28.

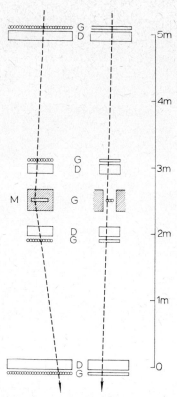

Fig. 27. Large, high resolution muon magnetic spectrograph, typical of a series of instruments developed by A. W. Wolfendale and his colleagues at Durham. On the left-hand side is shown a view of the instrument perpendicular to the deflecting magnetic field; the right-hand view is in the vertical plane containing the direction of the field. The spectrograph is triggered by a coincidence of five trays of Geiger tubes (G) (including a small guiding group of counters in the air-gap of the deflecting magnet (M)) which activates the four blocks (D) of high resolution detectors. The trajectories of deflected muons are determined from the detailed response of these blocks. (The construction of these blocks is better shown in fig. 28.) The extreme resolution of this instrument depends on the accuracy of track location in the (D) blocks and the long throw for which each track is traced. This last feature, and the limited aperture of the magnet air-gap imposes a much slower rate of data collection than in the contrasted instrument in fig. 28.

muon momentum from the angle is straightforward, and the other element necessary to obtain the spectrum, the collection efficiency of the instrument as a function of muon momentum is essentially geometrical. The primary spectrum in fig. 25 up to beyond 10^{14} eV is from such data together with detailed studies of muons by other

Fig. 28. A specialized muon spectrometer designed and operated by workers from Durham close to the central detector-station (A1) of the Haverah Park EAS installation (fig. 30). It is designed for detailed study of muons in very large showers, and some momentum resolution (as compared with the spectrograph in fig. 27) has been sacrificed in favour of increased collection geometry. It shows in detail the high resolution detector-blocks also used in the fig. 27 spectrograph. Each consists of a pack of many layers of small diameter electrodeless gas discharge tubes. These are activated from external electrode sheets by the operation of the shower array, and particle trajectories can be located through each detector-block to extreme accuracy.

methods and underground, and overlaps the region of direct observation of primaries; moreover the degree to which it is uncertain because of the factors (i), (ii) and (iii) introduced above is not very great.

An entirely different set of observations, those of N. L. Grigorov mounted in the 'Proton' series of satellites, are not easy to bring into full agreement with the derivation just described, and in fact suggest that the spectrum falls away rather more rapidly. That this discrepancy has not been resolved over a period of several years must be seen as emphasizing the difficulties of this whole area of measurement. We incline to regard the Durham spectrum as the more likely to be correct, since it does not seem to be subject to any uncertainty which does not also affect the 'Proton' experiments, while the exact status of measurements in an ' ionization calorimeter ' (the instrument used on these satellites) is not as clear-cut as the momentum spectrum of muons, which is one of the classical experiments of cosmic ray physics and which has developed through many inter-comparisons of work in various laboratories.

The region of the primary spectrum which follows, and which is shown in fig. 26 with the broken line, is of critical interest, because it is the site of the only positive feature which seems to occur over the whole eight decades of energy encompassed in this illustration, that in which the slope of the spectrum varies in a way which, even if it is not dramatic is certainly very much more rapid than in any other region between 10^{12} and 10^{20} eV. At the time of writing it is not possible to describe the course of this change nor even to establish how abruptly it actually takes place (over a factor of ten in energy or over a factor of a hundred). Indeed the basic feature that it is a region of rather abrupt change rests mainly in the confidence with which the adjacent regions of the spectrum are thought to be well-founded. Measurements of primary energy of any kind in the change-of-slope region are difficult and subject to criticism and an accepted form, in good detail, is not likely to emerge for some time to come.

4.6. The shower region

The formation of showers of secondary particles in the atmosphere, the way in which a shower grows to a maximum size and then dies away, was the main topic of Chapter 2. From primary energies of about 10^{16} eV upwards, the study of these showers, and the detailed problem of connecting observations from them with the primary particle from which they have developed, is the only approach which we have to estimates of the energy of primary particles and so of the energy spectrum to which they contribute. The reason why any method which involves the direct interception of the primary particle itself, or even of the extension of its trajectory through the atmosphere, fails is made very clear in fig. 26, for this region is characterized by an extremely low intensity of primary particles, extending from about 1 $m^{-2} y^{-1}$ near 10^{16} eV to something approaching 1 km^{-2} century^{-1} for the most energetic primaries yet detected.

The special value of the shower approach is derived from the way in which the secondary particles spread out from the ' axis ' (the continuation of the trajectory of the initiating primary particle) over comparatively large areas. The form of their distribution as a function of distance from the axis is regular, and because of this feature it is possible to gain the essential information about the primary particles carried in the secondary development from them by

Fig. 29. The core region of a shower of total energy about 10^{15} eV as recorded in a very large counter-controlled cloud chamber designed and operated by A. L. Hodson at Leeds. The chamber is mounted horizontally and viewed from vertically above, the area covered in the photograph is approximately $1\cdot6$ m $\times 1\cdot4$ m and the depth illuminated and photographed is about 30 cm. Tracks in a vertical shower would be seen more or less end on; the actual example shown is inclined about $30°$ to the vertical, and each individual track is passing down through the illuminated volume in the direction from top right. The important feature shown very clearly is the complexity and irregularity of shower particles within about 1 m of the true axis. There is probably still considerable nucleon cascade activity in this region.

sampling it at a limited number of points rather than having to observe it in detail all over.*

The relation between the intensity and distribution of secondary particles (the so-called ' lateral structure function ') and the energy of primary particles is at its simplest near to the maximum development of the secondary component, where the effects of fluctuations ((iii) above) are least important and where there are the greatest number of secondary particles to detect. Near sea-level the lowest-energy particles in the shower region (from rather below 10^{16} eV to perhaps 2×10^{17} eV) give rise to showers which are well beyond their maximum development, and really satisfactory measurements have not been achieved. So work at these energies has to take place at mountain altitudes, and of mountain stations that on Mt. Chacaltaya in the Andes near La Paz, the capital of Bolivia, is pre-eminent and has been for more than twenty years. Although at an extreme altitude for ground-based work, 5220 m (rather more than 17 000 ft), the installation at Chacaltaya is relatively easy to access (see page XII), and its facilities are open to workers from many countries.

Beyond about 10^{17} eV work can be successfully undertaken near sea-level, and important experiments are going on in many parts of the world, in the U.S. and the U.S.S.R., in Australia and in England. All of these involve semi-permanent installations spread over 10 km² or more and in operation for many years. All use significantly different detection arrangements and when results coming from these prove to be in substantial agreement, the very fact that different ' mixes ' of the various secondary components have been used adds to the overall confidence in the conclusions. Here we will take as an example of these the shower detector array at Haverah Park near Leeds.

The arrangement of detectors at Haverah Park is shown in fig. 30, where at each detector station the total area of detector (m²) is indicated. The 'sample' of shower particles is not a large one, perhaps 10^{-4}, but even this level of sampling is a good deal higher than that at the other large arrays. The relatively abundant particles of energy near 3×10^{17}–10^{18} eV are mainly studied using the closely spaced detector stations in the centre; for increasing energies the outer ring of detectors becomes more important. Showers from primaries of energy near 10^{20} eV are detectable over about 15 km², but for precision measurements and particularly for the determination of the spectrum, where exactly defined areas of collection must be used, even for the largest showers the condition is imposed that the axis must fall within the area defined by joining the outer ring of detectors to form a polygon.

* This regularity of lateral spread does not extend to the immediately axial region dominated by active nuclear particles. See fig. 29.

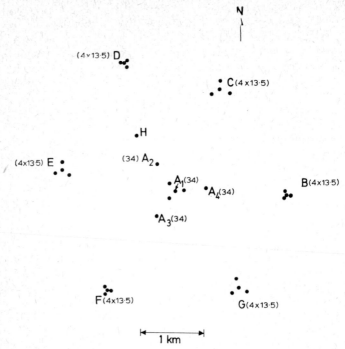

Fig. 30. The main detector array at the Haverah Park Extensive Shower system the composite detector at each point of which is made up of a grouping of unit detectors each of area about 2 m² (see text). The area of detector at each point (/m²) is given by the figure in brackets. There are also a considerable number of smaller (1 unit) detectors set out for specialized experiments, and several detector systems of quite different function for detailed studies of showers (see, e.g., fig. 28). For the main array of detectors, connections within the A-complex are by coaxial cable, connections between the central control (at A1) and the outlying sites (B . . . H) are by microwave links.

The actual detectors used at this array are built up of unit-detectors each about 2 m² in area and 1·2 m deep. They are tanks lined with white plastic and are filled with optically pure water, and what is actually recorded is the Čerenkov light* emitted by shower particles passing through the water. This light is diffusely reflected from the white lining and received at a photo-multiplier tube mounted in the lid of the tank. The signal obtained in this way is almost independent

* Named after its discoverer, P. A. Čerenkov, this light is the optical equivalent of a supersonic shock wave in the air. It is emitted when a charged particle passes through a transparent medium (here water) with a speed greater than the phase velocity of light in the medium. Since the refractive index of water is about 4/3, all relativistic charged particles emit Čerenkov light in water.

of the arrival direction of the showers and is fairly accurately proportional to the energy deposited by the shower particles in the water. Since this depth of water is about one eighth of an atmosphere in mass, these detectors have to be regarded as ' thick ' and indeed they absorb and measure a large fraction of the energy carried into them by shower particles.

The analysis of showers using data from this array is now performed using a very complex computer programme, and here we comment on only a few important points:

(i) Each record gives the time of occurrence of the shower, the intensity of signal at all detectors which respond and the relative times of arrival (to within about 50 ns) of the signal at each of the large (34 m²) detectors in the central part of the array. From these data, particularly as to timing, the local direction of arrival of the shower (zenith angle, azimuth angle) is determined, and so, since the time of the event is recorded, the celestial co-ordinates in right ascension and declination computed.

(ii) Over a period of years, study of the lateral distribution function for signals in the Haverah Park detectors has shown that only an extremely small change occurs with changing primary energy and the variation with zenith angle (derived in (i) above) also has been well-determined.

(iii) If each detector gave a precise measurement of shower signal density at its particular location, a simple computational programme could be written to determine the point of impact of the shower axis on the plane of the array, and exact values of the signal as a function of distance from the axis would follow. However, what is recorded is a *sample* based on only a limited number of particles, sometimes quite a small number. Such samples are intrinsically uncertain, and must certainly not be treated as though they are exact.

The real difficulty of the whole analysis stems from the modes available for treating the uncertainties of sampling, and the aim is to determine the most likely location of the shower axis and to give a figure of merit which describes the goodness of fit of the measured signals to a shower based on this axis and of the derived shower size. In fig. 31 the results emerging from this process are shown for a few showers of primary energy ranging from about 3×10^{17} eV to 3×10^{19} eV. The full lines show for each shower the lateral structure function at the level for which it gives the best fit overall, and the extent to which the individual measurements lie off these lines indicates the extent of the sampling problem. This illustration is for near-vertical showers (zenith angle less than 20°) and where for all of these, therefore, the same lateral form of distribution will apply.

75

What has been indicated in outline in the last section is the determination of a description of ground-level data for the shower, that is to say, of *what we can actually measure*, where the difficulties are those of any series of physical measurements as they arise from limitations

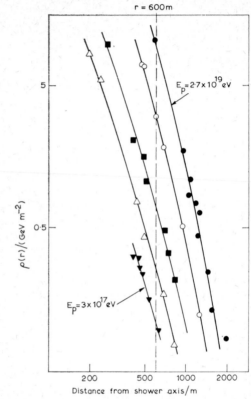

Fig. 31. Examples of best fit analyses of showers (Haverah Park data). All five showers are near-vertical ($\theta < 20°$) and may be represented by a lateral structure function of the same form. For each the most probable position of the axis is determined and then the signal at each detector is related to its distance from the axis, r, and is plotted at $\rho(r)$, the energy absorbed at the detector in GeV m^{-2}. The lateral structure function, for which the shape is known but the absolute intensity increases with energy, yielding the best fit is shown for each shower, and determines a quantity closely related to primary energy. The goodness of fit is shown by the way in which the measured values of detector signals fall around the lateral structure line. As explained in the text, the value of $\rho(r)$ at $r = 600$ m *both* falls within the spread of observations for the majority of showers *and* is relatively insensitive to the nature of the primary and to the details of early nuclear interactions. Accordingly $\rho(600)$ is an effective ground-determined quantity to be related simply to primary energy.

of equipment and observational technique. The next step, in which the ground-level data must be related to properties of the primary particles, involves arguments of a more difficult kind.

4.7. *Primary energy*

All simulations of showers lead towards the conclusion that near the maximum of shower development the size of the shower at the ground, however this has been determined, is more or less proportional to the primary energy, and this seems to be true of the different kinds of response at all the large shower arrays now operating. While the overall objective of shower studies is the determination of *both* the energy of primary particles *and* their nature, these two features cannot both be determined from a measurement of a single quantity. Accordingly, a first objective is to minimize the effects of the features to which the observations are not very sensitive, and to work out how the dominant factor, the primary energy, can be related to the ground-data in a way which as far as possible does not depend on assumptions about the nature of the primary particles nor upon details of nuclear physics in the early interactions.

The procedure to achieve this differs from one mode of shower detection to another and can be very complex. For the Haverah Park system upon which fig. 31 is based, the most effective measurement is identified as the density of deposition of energy in the sampling detectors at 600 m from the shower axis: this is the quantity $\rho(600)$ referred to in that diagram. The choice of this quantity is partly purely practical: as will be seen from fig. 31, it is a distance from the shower axis for which there are very often almost direct observations. Its more fundamental important quality arises because the signal observed in these particular detectors at 600 m from the axis arises approximately equally from the muon component and the electromagnetic component. Now the differences which come from varying assumptions about what the primary particle is, and about what happens in the early interactions, affect in particular the relative importance of the muonic and electromagnetic components. If we concentrate on this particular feature, changes in the relative proportion of the two components (as between one set of assumptions and another) will have least effect on the relation with primary energy for a ground-measurement in which the two components are known to be giving similar contributions, since to a first approximation any increase of one will balance a corresponding decrease in the other. This particular distance from the axis, 600 m, is shown as the thin broken line on fig. 31.

All the large-scale installations now operating yield data which are studied through these two stages. The first, which deals with direct observations and their interpretation on the ground is special to each

77

experiment, and will almost certainly be best understood by the people actually working with them. The second, relating ground data to primary energy, is always open to new influence by work going on anywhere about the nuclear processes which form the most unclear link in the chain. Uncertainty about these is the main factor which limits confidence in the translation from direct shower observations to values of primary energy. The high-energy end of the primary energy spectrum shown in fig. 26 (10^{17} eV upwards) is how this is to be seen at the time of writing. The index of the slope is probably good to within about 0·05, the absolute intensity to within perhaps 50%. This last figure is that which most of all reflects our limitation of understanding of very high energy interactions.

4.8. *Nature of high-energy primary particles*

Because the effect on shower phenomena at the ground of the nature of primary particles is small, and because it is inextricably related with the important early nuclear collisions, the details of which even for a proton primary remain uncertain, the problem of determining the nature of these particles is far more difficult than that of describing the total energy carried by the primary (irrespective of whether it is a proton, α-particle or a still heavier nucleus). At the time of writing there is no generally accepted view as to what the primary particles are.

What is coming to be agreed is that probably the most promising approach to finding out will depend on studies of fluctuations of shower development. Even from a proton primary every shower is different, and perhaps the most obvious starting point from which this difference grows is the distance into the atmosphere which this proton will penetrate before it makes its first collision. The average distance is usually supposed to be about 80 g cm^{-2} (although some evidence suggests a rather shorter distance), but this average must include many occasions on which the first interaction takes place at 20 or 30 g cm^{-2} and others on which it does not take place until 200 or 300 g cm^{-2}—that is to say a quarter of the whole atmosphere. Particularly when second interactions are also considered, this initial phase of shower growth is expected to lead to noticeable differences in detail of showers at the ground, some of which will, it is hoped, be measurable. Because heavier nuclei, even α-particles, are larger than protons they will make their first collisions after a shorter distance and the spread of this distance will also be smaller. Then, in the first two or three collisions, the heavier primary particle will tend to break up into individual nucleons each of which will then initiate an independent shower of its own. For these two reasons fluctuation effects will be expected to be smaller for heavy nuclei, and it may be that measurements will be possible which will yield some indication

78

whether those actually found to occur accord with proton-like primaries or heavy-particle-like primaries.

4.9. *Primary electrons*

The pioneer experiment of Thomas Johnson on the difference of cosmic ray intensity at balloon altitudes from the east and from the west established that practically all the particles incident upon the earth were positively charged, and left very little which might be ascribed either to a primary uncharged (electromagnetic) radiation or to primary electrons. It was not for many years after this that it was definitely established that there were electrons in the primary radiation, because as long as experiments were carried out within the atmosphere, even at the highest balloon altitudes where perhaps less than 1% of the atmosphere was above the apparatus, it was still difficult to distinguish for certain genuine primary electrons from the secondary electrons which were already beginning to be present as a result of normal electron–photon cascade development from particularly early nuclear collisions.

This area of cosmic ray studies was quickly advanced and, indeed, transformed, when experiments carried on satellites became possible, and over the energy range covered by fig. 22, where the electron spectrum is shown as a broken line it is now at least as well established as that of most positive nuclei.

The electron spectrum is now known as to general form up to energies $\sim 10^{12}$ eV. Below about 10^{10} eV, as for positive primaries, the spectrum is complicated: there is certainly strong solar modulation but very many electrons may actually be of solar origin, and a substantial proportion are certainly secondary to normal positive primaries, by ionizing collisions transferring large amounts of kinetic energy and by the ultimate decay products of pions.

In addition at very low energies (< 30 MeV), a most remarkable and unexpected situation has been shown to exist. The flux at these energies near the earth is of course susceptible to many influences, and these have often been clearly related to solar features. But there are well observed examples of increases of flux which it has not proved possible to relate to anything taking place at or from the sun. An explanation of these effects has had to await data from the Pioneer 10 spacecraft gathered during its close approach to Jupiter. Here discrete bursts of low-energy electrons several hundred times the normal quiet time electron flux were observed. These seem to originate in the magnetosphere of the planet, which must be the seat of the acceleration mechanism although the process is not understood. Electrons arising in this way have now been shown to contribute significantly to the (low energy) flux near the earth, particularly to

79

increases of flux when there are no solar features with which they can be connected.

From 10^{10} eV to 10^{12} eV, the limit of our present information, and, as for positive nuclei the range where local disturbances seem to become unimportant, the electron spectrum has an almost constant slope with a differential index of about $-3 \cdot 0$.   However this simple statement conceals uncomfortable uncertainties. Some workers conclude that the slope increases with increasing energy, others consider that there is no sign of change, the absolute intensities given at any particular energy differ among various workers by a factor of up to four!   Near 10^{10} eV about 1% of primaries are electrons and since the differential spectrum is certainly steeper than that for protons and other nuclei, the proportion of electrons becomes even smaller at higher energies.

A very obvious question is, why are there so few electrons relative to protons, α-particles and so on of comparable energy, and one is bound to relate this feature with the fact that electrons lose energy, during their passage through the interstellar region, by a process which does not operate for the heavy positively charged particles. This mode of energy loss is by the emission of radiation under acceleration and so is analogous to bremsstrahlung emission by electrons in the atmosphere.   However the acceleration involved is now that taking place in the magnetic fields of galactic space, the so-called ' synchrotron radiation '.

Synchrotron radiation is strongly energy-dependent, in fact:

$$ - \mathrm{d}E/\mathrm{d}t \sim E^2 m^{-4}, $$

where the dependence on mass, as was the case for particles within the atmosphere, is large enough to ensure that a process important for electrons can be insignificant for protons.   If electrons initially represented by a power law $E^{-\gamma}$ remained for long enough in such conditions, the effect would clearly have been more severe on the more energetic electrons, and the spectrum would move towards an equilibrium form $E^{-(\gamma+1)}$, together with an overall reduction of intensity.   ' Long enough ' in this context is a function of energy, and the higher part of the electron spectrum would be expected to show this steepening first.   With present estimates of the various quantities involved, the steepening would be expected to set in at energies near 100 GeV, and so the conflict of evidence about the form of the spectrum, as it stands at present, is important.

The radiation of energy from electrons in the interstellar fields lies largely at radio frequencies, and the galactic radio synchrotron emission which arises is measured in detail by radioastronomers.   What can be measured at the earth about cosmic ray electrons and about this radio-emission allows some details about interstellar fields to be studied.

4.10. *Solar particles*

It is a matter of convention whether we regard particles ejected from the sun with enough energy to be recorded by normal cosmic ray monitors on the earth as ' cosmic rays ' or not. The average intensity of these particles is very low indeed; their energy spectrum is much steeper than, and their mass composition quite unlike, that of the galactic cosmic rays. In spite of these differences, which indicate that solar particles are examples of an altogether different mode of production, for completeness we include a short account of them here.

Solar emission seems first to have been observed in February–March 1942, using detectors sensitive to the general cosmic ray intensity which registered increases of up to 10% in high latitude observations and small or negligible increases at equatorial stations. The increases only lasted a few hours. No further emissions were reported until July 1946 when very similar observations were reported. In November 1949 emission again took place, and this occasion was notable because for the first time the nucleonic component (which is most sensitive for low energy primaries) was recorded in a primitive neutron monitor at Manchester. The neutron-monitor showed a 600% increase, about fifty times that of the normal particle detector operating at the same time at Manchester. This immense difference, which is to be compared with a neutron-monitor/muon-telescope ratio of about two for modulation effects, indicates that the slope of the spectrum of these particles is far steeper than that of real (galactic) cosmic rays; another, and equally striking demonstration of this feature was shown at a more recent emission, when the *latitude variation* of the observed increase between Leeds and London (less than 300 km) was by a factor of about two!

The expectation in 1949 that such acts of emission might be expected every three or four years has not been maintained. There was a very strong outburst indeed in 1956 (described below), but since then effects at sea-level have been disappointingly slight, and it is tempting to connect this with the low level of solar activity in general in recent years. At the same time, however, the ability to observe on satellites for long periods beyond the atmosphere has revealed many small emissions at lower energies which are ineffective at sea-level. What has been observed in the half-dozen or so classical events are the rare occasions of extreme activity which have produced relatively high energy particles (that is to say particles of energy within the cosmic ray range and capable of producing effects at sea-level) in abundance.

The largest of these took place on 23rd February, 1956, at about 0400 h UT, when the counting rate of the Leeds neutron monitor increased about 5000% in much less than a quarter of an hour (fig. 32). From this sudden peak, the measured intensity decreased fairly regularly, and twelve hours later it was still far above normal rate.

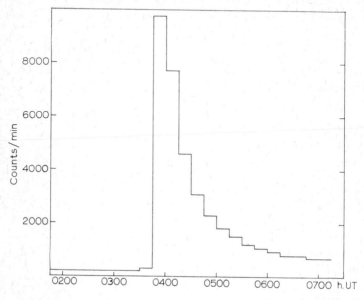

Fig. 32. Particles of solar origin observed at Leeds on 23rd February 1956 in a neutron monitor. The monitor was small, and the normal counting rate (left-hand side of graph) was about 200/min: it recorded at 15 minute intervals. The largest count in 15 minutes was at an average rate of nearly 10 000/min, and from the general shape of the curve there can be little doubt that the actual maximum rate was even greater than that average. Three hours after onset, the counting rate was still more than three times normal cosmic ray intensity.

A particularly interesting feature of this occasion could be followed because the whole effect was large and easily measurable for so long, and because already in 1956 there was a fairly good network of neutron monitor stations over the world. The monitor at Chicago lies at a similar geomagnetic latitude to that at Leeds (in fact slightly to the north of it), and when the response of these two monitors to solar particles was compared the result was very interesting. Between 0345 UT and 0400 UT, the measurement at Leeds was a maximum but that at Chicago was slight. For 0400–0415 UT, the effect at Chicago was a maximum but still smaller than that at Leeds by a factor of 1·8. This factor diminished in successive quarters of an hour until, for 0500–0515 UT, it reached a value of 0·72, and then within experimental uncertainty it remained steady for at least fifteen hours (fig. 33).

The interpretation of these observations is in accord with what has been discussed earlier relating to modulation of galactic particles.

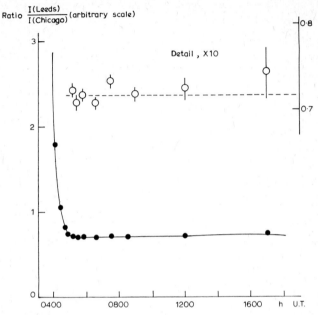

Fig. 33. Solar particle event of 23rd February 1956: ratio of proportional increases in neutron monitors at Leeds and Chicago.   Since the two monitors were not similar, the actual value of the ratio is not important. The graph shows that for about half an hour particles reached Leeds much more easily than Chicago; thereafter the ratio reached a steady value which remained unchanged to well within 10% for at least twelve hours.

For most of the first hour, when particles were arriving preferentially at Leeds, they were coming from the sun along a directed path and no large amount of diffusion was evident: because of the geomagnetic field these particles swung round to reach the earth at vertical incidence on the early-morning side of the earth rather than the late-evening one.   More recent studies clearly indicate that the particles came to the earth directly along field lines from sun to near the earth. Within an hour this injection of new particles into the neighbourhood of the earth became unimportant, and for the best part of a day particles diffusing isotropically were detectable near the earth. This figure, more than any other, perhaps, illustrates how efficient the diffusing mechanism is in slowing down the overall movement of charged particles in a scattering medium, here the solar wind, to something far less than the actual velocities of individual particles.

While many theories have been proposed for the mechanism of acceleration of solar material in these events, our picture of what is happening in detail is still very uncertain. If nearby and relatively simple happenings present such difficult problems, it is understandable that discussions about the sources and origin of real cosmic rays can only be undertaken along the broadest of lines.

4.11. *The carbon-14 dating problem*

A quite unexpected example of the interaction of scientific disciplines has led in the last few years to evidence of changes in cosmic ray intensity which, although enough is not known about them to see quite where they should fall in the plan of this book, are most plausibly thought of as modulation phenomena. About 1949 Willard Libby pioneered the application of measurements of the proportion of the radioactive carbon isotope, ^{14}C, to normal ^{12}C in ancient organic remains (particularly in wood) to the determination of the age of samples taken from the material being investigated. This isotope has a half-life of rather less than 6000 years, and so, if there were no difficulties in the application of the method, age measurements dating back perhaps 20 000 years or more would be possible. The presence of ^{14}C upon the earth depends upon the capture of slow neutrons, released at cosmic ray interactions in the high atmosphere, by nitrogen nuclei in the process:

$$^{14}N + n \rightarrow {^{14}C} + p,$$

and the dating application was based on the assumption that cosmic ray intensity had remained constant over the time scale for which the method was to be used. Since the life-time of the isotope is so long, it is safe to assume that the ^{14}C, formed mainly at very high altitudes, gets thoroughly mixed in the atmosphere as a whole and in soluble material in the accessible surface layers of the land, and that it thus enters in constant proportion into the carbon content of all currently living matter. That there should be no exchange of carbon atoms between this material, when dead, and the surroundings over a period of thousands of years is not so obvious, but does in fact seem to be substantially true.

The method was applied widely, and a ^{14}C time scale was developed relating to archaeological material: there were striking successes with the ^{14}C data agreeing closely with that of some of the relatively infrequent examples of ancient wood for which a prehistoric dating was well established by other means. However, as time went on rather large areas of discrepancy arose which could not be dismissed either as experimental errors or as casual misinterpretation of individual items of archaeological evidence. The archaeologists' scale and the ^{14}C scale differed and there was something of a deadlock. From the point of view of archaeologists, the problem was serious, since the relationship in time of evolving cultures in different parts of the world was basic for some of their most interesting problems.

What has now clarified the situation is the discovery of wood still in the form of the original tree trunks which can by study of the annual growth rings, and matching rings in trees of differing degrees of antiquity, provide as it were a digital scale of years back about 8000 years. An at first sight most unprepossessing species of tree, the California bristlecone pine, some specimens alive and some dead, but *in situ*, has now provided what seems to be an absolute time scale against which the prevailing ^{14}C concentrations, and by implication, the cosmic ray intensity, over the last few thousand years can be examined. The evidence seems strong that variations have taken place; it manifestly does not indicate whether these variations arose from local modulation or from a more widespread change in galactic intensity.

The likelihood is strong that the changes revealed here are local to the solar system. It is true that they cannot be linked directly with features of which we have direct measurements, for these only spread over a few tens of years, whilst the finest variations which are yet apparent from the ^{14}C calibration spread over several centuries, and the important ' drift ' which took archaeological and ^{14}C datings drastically apart seems to have been taken place during something like 3000 years (4000 B.C.–1000 B.C.). These features will certainly be increasingly well established, with further fine detail emerging, but for the moment any variation in this order of time scale is much more probably a solar or terrestrial variation than anything from outside. To notice only one feature, even at sea-level the differences of the 11-year variations of intensity during the last two solar cycles has been striking, but high in the atmosphere, where the main ^{14}C production takes place, it has been very much more dramatic, both as regards modulation and as regards the injection of solar particles. We do not know enough about remote solar cycles to exclude the possibility of periods when the average 11-year cycles over hundreds of years were vastly different from the few for which we have good information. Another possibility relates to the variation of the geomagnetic field. Studies of rock magnetism, which refer to far longer time scales than that covered by ^{14}C, undoubtedly indicate major changes of the geomagnetic field and so of the extent to which it allows the lower energy cosmic ray particles to reach the earth.

More recent but less intense changes could have led to significant changes in the production of ^{14}C. For these reasons, solar and terrestrial variations leading to the sort of change of ^{14}C formation required to understand the data as now presented do not appear at all unlikely, but it is interesting to speculate when (if ever) this particular variation would have been discovered were it not for the pursuit in the first place of the quite different objective of archaeological dating.

CHAPTER 5
cosmology of cosmic rays: background

5.1. *Introduction*

THE great value of recent advances in cosmic ray physics depends upon the way in which these have interrelated with other aspects of cosmology, and so we can hardly avoid prefacing our speculations about them by summarizing the sort of environment in which we think of cosmic ray particles originating and travelling. The unit from which we can develop this study, both inwards and outwards, is a galaxy, and specially our own particular galaxy, which is the one to which we always refer unless some different context is clear.

Seen from outside, our own galaxy would appear as a disc of radius several times 10^{20} metres, with a thickened and certainly very dense central hub and with the less dense parts of the disc, further from the axis, lying in roughly spiral arms which are the product of the rotation of the whole system about a central axis normal to the plane of the disc. This description follows the fact that all measurements upon our own galaxy seem to confirm that it is very similar indeed to the nearest of the large external galaxies (the great Andromeda nebula) which shows exactly such a structure; moreover studies by radio-astronomers of the distribution of hydrogen in the galaxy, as it is detected using the 21 cm emission line, confirms this general form. By no means all external galaxies appear to be of this type, but very many are spiral systems. In fig. 34 we show a photograph of the galaxy M101, which is particularly well seen from our own galaxy because we are looking more or less along the rotation axis of the spiral. Our own galaxy, like Andromeda, probably has a much more substantial hub and its spiral arms do not spread out so quickly, but the very fact that differences of detail of this kind are observed and classified is only possible because there are so many spirals even within the moderate distances for which it is possible to see their structure. The environment of our own sun and the earth are probably not very different from that of a similar star in M101, for they are well out along one of the arms and in a position from which other arms, both inside and outside their own can be recognized.

The obvious disc and hub are prominent because the great majority of stars in the galaxy lie in them, but there are features which suggest that the whole of a galaxy is better thought of as thicker, perhaps even spherical rather than disc-like, with the disc forming an equatorial plane. The remaining part is only thinly populated with stars, and most of these are to be found in relatively close groupings, the so-called ' globular clusters '. The fact also that this volume, which is sometimes referred to as the ' halo ' seems possibly to be the source of a detectable radio signal, suggests that the interstellar gas extends into it at a considerably greater density than is found right beyond the confines of the galaxy.

Fig. 34. The galaxy M101 photographed from the earth: it is chosen for illustration because we happen to be fairly near to the spin-axis of its spiral, and so have a ' full-face ' view. Differences between M101 and our own galaxy are mentioned in the text, but the general similarity is strong. Some of the dark features of the photograph, not far outside the central hub, which emphasize the spiral structure are obscuring dust clouds rather than apertures right through the galactic disc, but their conformation to the spiral structure is as significant as that of the distribution of stars which gives rise to the bright sectors of the arms.

5.2. The ' big bang '

Looking outward from our own galaxy it is certain that the immense number of other galaxies that can be seen are neither evenly spread about nor all similar in form. Their variety probably represents both differing stages of evolution and also evolution along different paths. Their irregular distribution through the volume of space is shown by the fact that our galaxy, together with the Andromeda galaxy, are two prominent members of a local group of galaxies. This is not a very big group, but within the last year or two further members of the group comparable in size to our own and the Andromeda galaxy have been detected, much obscured by interstellar dust. An exceptionally notable grouping of galaxies very close to our own, and therefore prominent to us, is that to be seen in the constellation Virgo, which appears to involve some 3000 individual large galaxies. Photographs of parts of the ' Virgo Cluster ' show dozens, perhaps hundreds, of galaxies, many of which must be like our own.

One of the best known features of the great system of galaxies as a whole is that they appear to be receding from each other and that as observed from any particular point, for example the earth, this recession seems to be proportional to distance. This conclusion is drawn from observations of the ' red shift ' of identified spectral lines, in which these lines are found to be displaced towards the red end of the spectrum by an amount proportional to the distance away, a feature which is interpreted in terms of the Doppler effect as arising from a relative motion in the line of sight of the source and the observer with a velocity proportional to the distance away. A consequence of this situation is that in time past the whole of the system of observable galaxies which is found to have this characteristic must have been located within a much smaller volume than that which it now occupies, since the motion is, and must have been, one of general expansion. The argument may be carried through to fix a time at which this volume was very small.

This expansion from a small volume forms the basis of what is the most widely accepted model for the immediate origin of the universe of galaxies as we now see it. It is supposed that at this zero of time the whole of known matter was concentrated in one large nucleus-like structure. A detailed understanding of this object cannot be expected, nor anything about its earliest history, but it seems to have exploded with an ejection of matter which has spread out to the form which we can now observe. This is the ' big bang ' theory of the origin of the universe, and, if it is substantially correct, this theory has a further consequence which will relate very closely to cosmic radiation studies. But before we can examine this, it is worth digressing to emphasize two particular points.

In the first place, no measurements we can make today can establish the paths of existing galaxies, traced backwards, to an accuracy that would bring them into so small a volume at that time. There is a very big gap between what we can observe and the hypothesis which is here drawn from it. Secondly, even if this model is precise as to what happened from the instant at which an intense concentration of matter existed, it is not an account of the creation of the universe, and it does not pretend to offer any view as to how this object got into such an unstable configuration. What it does indicate, however, is that the moment of this explosion represents a situation beyond which it is very unlikely that any investigations we make now will ever tell us anything about earlier states of the universe: it was an episode which effectively blocks information from preceding epochs (if any) affecting observations undertaken today. It was, if you like, the moment from which the ordered world which we know assumed that order, and one ought to be reluctant to read anything more into this particular cosmological postulate.

But returning to our own time, were this ' big bang ' postulate correct, there is another component which ought now to be present in space and detectable as well as the multitude of receding galaxies. In the first instant of explosion, whatever the composition of the starting material, the initial expanding concentration would consist both of matter (in the form of particles) and of radiation, and for some time at least the concentration of both would be so high that energy would be exchanged easily between them, and there would be something near an equilibrium state. It is not difficult to see that this high efficiency of interaction cannot have continued indefinitely, for nobody would argue that at the present time matter and radiation are interacting efficiently, when individual light quanta travel for millions of years from a source before a few of them are intercepted by the earth and provide us with information about whatever it is from which they have come. What is not easily identified is the particular moment at which the transition from efficient energy exchange to inefficient energy exchange, that is to say, from equilibrium development to separate development, can be said to have taken place. However, calculations attempting to fix this time agree in putting it very early in the history of the universe as we know it now, and from then onwards matter and radiation have spread out almost independently.

The expansion of matter has followed a complex history, and, perhaps even from the instant of the ' big bang ', there have been definite irregularities in its structure which have determined how clusters of galaxies, then galaxies and then individual stars have separated out into the distribution we now know. What is certain is that relatively inconspicuous irregularities at an early stage can have set a pattern which has been intensified by gravitational effects, in which dense regions have become still more dense at the expense of the less dense regions beside them.

For radiation this concentrating effect does not happen, and we would expect a much more even distribution of radiation still persisting. What this radiation will be like is a much more difficult problem; the answer will come out in terms of particular models of the universe, and these models are not easily understandable by anyone but the cosmologists who specialize in that sort of theorizing. It is probably easiest to consider the problem as follows, although many readers will certainly be able to see some of the difficulties. We suppose that the radiation now in our own region of the universe occupies a certain volume isotropically and, at the time of decoupling (that is, the transition to separate development) between matter and radiation, also occupied a volume which, of course, was then much smaller. If this is so, we can imagine a small part of it inside an imaginary box with totally reflecting walls, for if these walls were removed, what would then enter would be exactly like what was reflected when the walls were in position. So, treated in this way, the problem is one of the cavity radiation (black-body radiation) which characterized the earlier interacting phase, now decoupled and imagined as in a perfectly reflecting box which is being all the time expanded in size. Under these conditions the total energy of radiation in the box remains unchanged, there is no interaction between the radiation and the walls of the enclosure and its effective temperature falls. Followed through to the present time, this leads to the conclusion that by now the radiation will be very cold indeed, and estimates of this temperature all fall somewhere about 3 K. Its intensity and spectrum will be those appropriate to cavity radiation at this temperature.

If such primeval radiation exists, it will be found to have a significant interaction with cosmic radiation, and we discuss later the extent to which this interaction is or should be detectable in our measurements. So here we identify a point of contact at which cosmic radiation meets another aspect of cosmology, and where their respective conclusions have to be reconciled. Information about primeval 3 K radiation can be looked for by other means than by its effect on cosmic radiation: some direct observations are consistent with the postulate of such radiation, but up to now these have explored only a very limited part of the spectrum of the radiation, have only roughly shown it to be isotropic near the earth and have offered no evidence at all that it extends far beyond our own galaxy.

We have already mentioned that the galaxies which we can observe do not all look alike, and are not all of the form of our own and of the Andromeda galaxy. The differences are emphasized when we begin to explore galaxies using other and newer techniques as well as simple optical observation. In particular, some appear to be sources of enormous intensity in, for example, radio-emission and X-ray emission and are evidently in a state in which (perhaps fortunately for us) no nearby concentration of matter exists at present. These objects, like the low temperature radiation, are interesting to us, for if we are forced to the conclusion that some cosmic radiation has its origin outside our galaxy it is not plausible to think of it as coming from galaxies no more active than our own, which (as occurs, of course, for star-light) would then swamp all others as a source of this sort of cosmic radiation. We would have to look hopefully towards these super-energetic objects which are the most obvious power houses of the observable universe.

5.3. *The local galaxy*

While we of course speculate about the relationship of cosmic radiation with the greater universe beyond our own galaxy, and we have already discussed the short-range modulation which it undergoes inside the solar system, and as a direct result of conditions at the surface of the sun, its relationship with our own galaxy as a whole is bound to be particularly important, and in making our inward-looking survey of our galaxy we will slant it towards cosmic ray problems.

We have already recorded that the galaxy is several times 10^{20} m in diameter. Its mass is known with considerable accuracy and is about 5×10^{35} kg and its mean density is about 10^{-20} kg m^{-3} or about 10^{-23} g cm^{-3}. A great deal of the mass of the galaxy is in the form of large condensed objects (stars and planets): the density of the interstellar gas is only about 10^{-24} g cm^{-3} and the density of dust considerably lower even than this. Regarded simply as matter with which cosmic ray primaries might collide, it follows that the galaxy is relatively transparent. If we consider only the gas and the dust, the whole diameter of the galaxy corresponds only to about $(10^{22} \times 10^{-24})$ g cm^{-2}, or at most ten times this, which is very much shorter than the collision range of all species of nuclei, while if we bring in the remaining mass of the condensed bodies, almost all of this is collected into relatively small volumes which are immensely and wastefully over-thick as particle absorbers. Most of the condensed matter is present as stars not unlike the sun, but even the earth, which is on a much smaller scale (and therefore not so obviously over-thick as an absorber) presents something like 10^8 collision path lengths to cosmic ray primaries. The ineffective distribution of the greater part of matter in the galaxy therefore means that it is, as a whole, and as regards collisions of cosmic ray primaries with its general distribution of atomic nuclei, quite transparent.

But the most important feature of the interstellar gas is that it is highly ionized, forming a plasma which is, relative to its mass, an extremely good conductor of electricity. The result is that magnetic fields, however they have been produced, remain attached to the same sections of interstellar plasma as these move about for very long periods of time. What may be unexpected here arises as a matter of scale. In the laboratory one comes across eddy currents, which delay the decay of magnetic fields for a second or so: it is not easy to jump, in imagination, from apparatus of dimensions like 10 cm to interstellar clouds perhaps 10^{15} times as big, and to the decay periods of many millions of years which are appropriate to magnetic fields in these great volumes. But these field distributions are the most important property of galactic structure for many features of cosmic radiation.

As to the origin of these fields, obviously those existing now in the interstellar gas had their origin so long ago that we have no record about them, but we may try to visualize their development from examples of stronger fields which can now be seen, and which will, long in the future, have reverted to a less spectacular condition. We have already discussed the fields carried out from the sun on the solar wind, which originate from active centres in the surface of the sun, and which endure at most for a few years in the separate bubble of solar gas (the solar wind) which is encased within the interstellar gas. There is also strong evidence of fields, often very intense, connected with most of the obvious centres of activity in the galaxy, for example supernovae, pulsars, neutron stars and so on.

On the whole, strong fields are related to above-average dense gas clouds. In fact there is an inter-relation here, with the field unable to break away from, or decay in, the plasma of the gas, and at the same time holding together a relatively dense cloud of gas which would otherwise disperse into the surrounding less dense region.

Radio-astronomers can detect, relative to ourselves, dense clouds which consist mainly of hydrogen, by means of observation of the 21 cm hydrogen line. Recently, using the Zeeman effect of this line, the general magnetic field within individual clouds has been estimated. The cloud immediately surrounding us carries only a rather weak field, but another in the galactic arm outside our own has a much larger one which has been measured to be of strength about 2×10^{-5} gauss $(2 \times 10^{-9}$ T).

The interstellar magnetic field is clearly a very complicated one, both in magnitude and direction and has obvious resemblances to that entrapped in solar wind. In some regions, particularly in the spiral arms of the galaxy, we would expect a background degree of order in the field,* but even here it is to be expected that there will be fairly strong discontinuities. Essentially this system of fields provides a scattering environment comparable to that in the solar wind close to the earth and we must expect very similar questions to arise and similar conclusions develop about transmission of cosmic ray primaries in interstellar space as we did in interplanetary space, although on a quite different dimensional scale.

In discussing the situation within the solar wind we more or less took it for granted that the mass thickness of the interplanetary gas was negligible for collision purposes, and a simple calculation (assuming for example that the solar wind spreads out to 10 times the

---

* Radio synchrotron emission (p. 80) is polarized in a way which supports this conclusion.

distance of the earth from the sun and that it contains as a maximum perhaps 100 atoms per cubic centimetre) probably satisfied the reader that this was a very safe simplification. In the interstellar system this neglect of particle collision is not so well based, although we have already shown, in a rough sort of way, that a particle traversing the whole galaxy in a straight line would have only a very slight collision probability. But we have not in so doing disposed of the possibility that cosmic ray primaries, if they follow diffusive and therefore potentially much longer paths, may not be the subjects of a significant degree of collision resulting in one form or another of nuclear disintegration.

The magnitude of magnetic fields, required in interstellar space in order that cosmic ray electrons shall emit the intensity of galactic radio synchrotron radiation which is observed, ensures that almost all cosmic ray particles are moving in a strongly diffusive medium which, as was observed locally and in detail for solar particles in February 1956 (p. 83), will effectively retain particles for long periods in limited volumes near to their sources, the effective volume increasing with increasing energy. It is within the limitation set by this diffusion that one has to speculate about ' sources '.

5.4. *The Fermi mechanism*

For a time, some years ago, it seemed possible that there would be no need to postulate any specialized source regions, because, as the great Italian physicist Enrico Fermi first pointed out, it is possible for charged particles, such as the primary cosmic rays, to gain energy in the simple act of diffusing between scattering centres moving in the interstellar gas.

In fig. 35 we show a simplified one-dimensional illustration of this rather surprising proposal. Two scattering centres represented here by the surfaces of regions of strong magnetic field are shown in one case approaching each other, and in the second receding. Between them a cosmic ray particle bounces backwards and forwards. If this motion were not relativistic, one could see straight off that each time the cosmic ray particle collided with a field region moving inwards in the diagram, it would gain energy, while equally, if the field region was moving out of the diagram, it would lose energy. This feature remains, even when the particle is relativistic. Now, if two scattering regions are moving towards each other the time interval between successive reflections is shorter than if they are moving apart, so the gain of energy by the particle in (*a*) per unit time is greater than the loss of energy in (*b*).

The situation in this example is, of course, very highly idealized, but the principle is not effectively changed in a chaotic three-dimensional system of scattering centres, when approaching centres are as likely to be encountered as receding ones. In such a situation, the cosmic ray particle will, overall, gain energy. Perhaps the most vivid way of visualizing this is to accept that here we are thinking of something very like a familiar problem in the kinetic theory of gases. The clouds of denser gas with strong magnetic fields form, as it were, large molecules in a kinetic theory situation. Into it much smaller molecules, the cosmic ray particles, are injected and the whole system tends towards a steady state of equipartition of energy! If this process could go on long enough, the limit of energy which might be gained by the cosmic-ray particles would be far beyond our most extreme expectations.

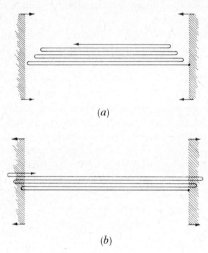

(a)

(b)

Fig. 35. The Fermi acceleration mechanism. A one-dimensional simplified illustration of a particle (a) gaining energy by repeated reflections between approaching scattering regions and (b) losing energy between receding scattering regions.

Unfortunately, this very attractive proposal turns out to have fatal weaknesses, and the most decisive of these is that it is far too slow. We know, for example from the studies of 21 cm Doppler shifts for hydrogen clouds, the number and relative velocity of the gas clouds which we picture as acting as scattering centres for quite a large section of the galaxy in our own neighbourhood, and from these it is straightforward to calculate the rate at which particles might gain energy. It turns out to be so slow that a quite negligible gain will take place before the particles have covered a total path many times their range for nuclear collision. The consequences of these nuclear collisions are complicated, but in fact they have a great deal in common with those of a primary particle when it first enters our atmosphere.

They involve the fragmentation of large nuclei, and for all, whether large or small, the transfer of a significant fraction of the energy into a meson component which, in the interstellar environment, will always degrade to electrons and photons.

Although the Fermi mechanism as a general property of interstellar propagation fails insofar as it does not seem able to offer an important means of accelerating particles, it does direct our attention to regions which may be quite different, and capable of being seriously thought of as sources; regions in which the conditions for this process are much more favourable, with the relative numbers and velocities of scattering centres very much greater, and where they have entrained in them far more intense magnetic fields. Such conditions do exist, for example in the debris left after the explosion of a supernova. However, even here it is by no means certain that this mode of acceleration will take place, for there are other means of accelerating particles in these regions which are at least as likely. Moreover these tend to be objects where matter is expanding away from some cataclysm in the past, and the argument which we followed in illustrating the Fermi mechanism postulated a steady state situation with collisions taking place with scattering centres which were as likely to be approaching as receding. If the whole object is something that is expanding, the situation is not like this, and an excess of collisions will be at receding scattering centres. Indeed, an ' inverse Fermi effect ' is sometimes discussed under these conditions, in which the tendency will be for particles on average to lose rather than gain energy.

We have finally to say something about the chemical composition of the galaxy. There are obvious reasons why the earth is not a typical sample. The light elements, hydrogen and helium in the free state will have had gas-kinetic velocities in the atmosphere which have allowed them to escape, and so hydrogen, except insofar as it exists in chemical combination, and helium almost altogether will have been lost. The same sort of conclusion may apply also to other volatile elements. When we examine stars, and the sun in particular, the problem becomes complicated, and it appears that there are at least two variables, the composition of the star when it first condensed into a recognizable entity and the stage of development which it has reached in its individual existence. At the present time, as regards the sun, very little change of composition is taking place except the conversion of hydrogen to helium, which is the essential process by which solar energy is now being released. The heavier elements in the sun, the existence of which has been known for a long time by the identification of the Fraunhofer absorption lines in the surface spectrum, seem to have been present in the original material at the time of its condensation.

However, there are stars, and good examples are those in the globular clusters which are found in the galactic halo, which are very poor indeed in heavy elements, and which are almost pure hydrogen–helium stars.

A tentative approach to this difference is to distinguish between first and second generation stars. Both condense with a composition determined by the gas cloud from which they have been derived. If the cloud is substantially pure hydrogen–helium, a star poor in heavy elements results. If the cloud, on the other hand, has a considerable mixture of the heavier elements, a star condensing from it will also have such a composition.

Why should gas clouds differ so much in composition? Whatever the composition of the gas before any galactic condensation began we can say something about that which is produced under certain circumstances of stellar evolution. The history of the life of a star is governed more than any other factor by its mass, the more massive it is the quicker and brighter is its normal life, while the conclusion of this, whether quiet or explosively violent, is extremely sensitive to mass. Violent episodes in the life of a star involve extreme conditions of temperature and density under which heavier elements are synthesized from the basic lighter material. Thus second generation matter, that which has come from the climax of the life cycle of a star, and which is then exploded into the surrounding space, tends to be rich in heavy elements. We may plausibly suppose that first generation matter, which has not yet experienced this transformation, is poor in and perhaps even free from anything beyond hydrogen and helium. Such a picture is not inconsistent with a ' big bang ' origin which on most models would yield essentially a pretty pure hydrogen–helium mixture in proportions which are compatible with what we believe to exist in interstellar space.

For cosmic ray studies we are concerned with two aspects of the abundance of the elements in the galaxy. The interstellar gas is the medium through which the particles move, and its composition determines in detail the possibility that nuclear collisions of any particular primaries will take place, and what happens in these collisions. On the other hand, the unusual and eruptive regions are probably those at which the acceleration of particles to cosmic ray energies takes place, and it is the composition here that is going to determine the nature of the material which constitutes actual primary cosmic ray particles at source.

CHAPTER 6
cosmic rays in the galaxy

6.1. *Introduction*
IT has been known for a long time that the identification of ' sources ' of cosmic rays is not at all straightforward, and this difficulty has been emphasized by the description of our galaxy, and its potentiality as a scattering environment, given in the previous chapter. Even the most general question: are cosmic rays of galactic origin and mainly contained in them, or are they a general feature, more or less freely moving in and out of galaxies, is not capable of a clear answer. In what follows, we shall make the *assumption* that at low energies, say up to 10^{12} eV and rather beyond, what we observe are certainly local cosmic rays, special to our own galaxy, while at the highest energies yet observed ($\sim 10^{20}$ eV) the particles are likely to be of the second kind, moving among galaxies, and in and out of them. We reach this final conclusion rather by default, for it seems very hard to put forward a mechanism to retain these extreme energy particles inside our galaxy, as far as we know it, with any degree of conviction.

From this standpoint of almost all cosmic rays being galactic and a very few of great energy reaching us from outside, the main features on which we have to build are:

  (i) the energy balance implied by the known intensity of cosmic ray particles,
 (ii) the conclusions coming from detailed examination of the charge spectrum of primary particles,
(iii) the isotropy of cosmic ray particles reaching the earth,
(iv) the change of slope of the energy spectrum near 10^{15} eV (fig. 26, p. 66.)

6.2. *Energy considerations about galactic cosmic rays*
The differential energy spectrum of primary particles at the earth can be written in the form:

$$I_{\mathrm{d}}(E)\mathrm{d}E \sim E^{-(\gamma+1)}\mathrm{d}E;$$

part of this spectrum is shown in fig. 22 and its extreme extent can be visualized from the integral spectrum of fig. 26. These show straight away that, except in the very lowest region of energy, the greater the particle energy, the lower the differential intensity. Indeed, even in the lowest energies shown in fig. 22 this would probably be true outside the solar enclosure, although it is not true inside it and near the earth. Our first aim is to determine in what part of the spectrum most of the energy of cosmic radiation is to be found. From this differential spectrum we can write the amount of energy carried by particles of energy greater than, say, E_0 in the form:

$$\text{energy } (>E_0) \sim \int_{E_0}^{\infty} EI_\text{d}(E)\text{d}E \text{ or } \int_{E_0}^{\infty} E^{-\gamma}\text{d}E.$$

If the index, γ, were a constant for all energies, this expression would show immediately that γ must be greater than unity, because if it were not the total energy would be infinite. Inspection of the graphs given earlier show that γ is not a constant, but varies over the whole spectrum, although for large ranges of energy and particularly at the higher energies it does not vary much. At very low energies and close to the earth it is in fact less than unity, but from 10^{10} eV upwards it seems always to be greater, and this is satisfactorily in accordance with the physical necessity that the energy carried should be finite. In particular, from $\sim 10^{12}$ eV upwards γ is at least two: if it were exactly two, we would then have:

$$\sum E_\text{p}(>10^{12}\text{ eV}) = 10 \times \sum E_\text{p}(>10^{13}\text{ eV}) = 100 \times \sum E_\text{p}(>10^{14}\text{ eV}) \ldots$$

and so on, where $\sum E_\text{p}$ represents the total energy reaching the earth carried by primary particles of the energy indicated in each expression.

What this establishes is that almost all of the cosmic ray energy is carried by primary particles of relatively low individual energies, and in particular by particles which lie in a well studied region over which the spectra of various nuclei can be described confidently. From the data available, then, an order of magnitude of energy can be identified around which the main energy-carrying component of the primaries lie, and this turns out to be about 10^{10} eV/nucleon.

Assuming that the life expectation of these particles from the time of their formation is a few million years (see pp. 104, 122) we can now go on to estimate how much energy is involved and the power input required to maintain it.

Integrating the energy content over the spectra already given (fig. 22) yields the result that near to the solar system the total kinetic energy of cosmic ray primaries is about 1 MeV m^{-3}, perhaps rather more. If we are placed in a reasonably representative part of the galactic disc, the disc of volume $\sim 10^{60}$ m^3 has an energy content as kinetic energy of cosmic rays of:

$$\sim 10^{66} \text{ eV} \sim 10^{47} \text{ J},$$

and this energy has to be replenished once every few million years, say once every 10^{14} seconds. Therefore, *assuming that the cosmic ray population which we observe does represent a steady state*, it requires to maintain it an energy input at the rate of 10^{33} W!

The situation with regard to electrons is not dissimilar. The lifetime of electrons is certainly shorter, but we are able to quantify the modes of loss of energy with some confidence. The total intensity of galactic radio-emission (synchrotron radiation) is about 3×10^{31} W and loss in Compton collision is rather greater. The energy replenishment of this component is accordingly about 10^{32} W, say one tenth of that required for the positive particle flux.

We have no means of actually testing whether the cosmic ray regime which we know is a steady state, but we can safely say that unless it were mostly the product of a single near and recent event it would on a cosmological scale be most unlikely to exhibit any change in so short a time as a few million years. Any speculation which treats cosmic rays as a normal feature of a galaxy such as our own must provide energy at something like this rate.

This estimate refers specifically to the galactic disc and does not pretend to include figures about the galactic centre, nor about the galactic halo. The first of these might indeed be a net source of energetic particles and of energy relative to the disc: but if it actually is a source, then the particles coming from it would be much more abundant in the near parts of the disc close to the centre, and will not have been taken into account in our calculations. As to the halo, its volume is not likely to be more than a hundred times that of the disc, the particle density in it is probably lower and the lifetime of particles much longer: it can hardly make great demands on the sources of energy.

6.3. *Sources of energy*

It is now possible to offer some sort of specification for source conditions, within the galaxy and referring to the lower primary energies:

  (i) Sources must exist which, in spite of strong diffusion, are effective in our own part of the galaxy, and if this is a typical region, similar ones must operate throughout the galactic disc.

 (ii) They must be regions in which nuclei of the whole known range of composition are related with very large releases of energy from which individual particles can attain energies at least up to amounts like 10^{13} to 10^{15} eV. (It is legitimate to postulate that very many of the normal cosmic ray sources do not emit the most energetic sort of primary with energy $> 10^{15}$ eV.)

(iii) Over the whole galaxy the energy output from these sources must be at something like the rate indicated above, that is to say at about 10^{33} W, as regards kinetic energy of cosmic ray particles alone, and so in total, to include other modes of dissipating energy, probably at the very least ten or a hundred times as much, since one cannot seriously imagine a natural system highly efficient for the production of cosmic ray particles and yet inefficient for the release of energy by other quite straightforward processes.

For many years much the most plausible candidates for a place in the inventory of sources have been the so-called ' supernovae '. These are rare; there has not been one which has appeared as a striking object as seen from the earth for more than 300 years. The energy release in them is very large indeed and one may expect that it is in conditions such as these that individual particles can be accelerated to the ' cosmic ray ' energies which are so much greater than those required for all other known phenomena. But one has to stress how little we know in the way of direct observation of the early (and doubtless most striking) features of supernova eruptions.

The Soviet physicist, V. L. Ginzburg as long ago as 1955 examined seriously the possibility that supernova eruptions might play a major part in the production of cosmic rays from the point of view of being able to provide for an input of energy of the right order. At that time it was already realized that the matter involved in these eruptions would be relatively rich in heavy elements, as the experimental evidence about the composition of cosmic ray primaries would require.

It is for the electron component that some degree of quantitative detail can now be added to the supernova source hypothesis. In what is surely the best known supernova, that which was seen to erupt in the constellation of Taurus in A.D. 1054, now the Crab nebula, the relativistic electron content over the lifetime of the object can be estimated from the very characteristic synchrotron radiation now observable. Ginzburg estimates this electron component as normally carrying perhaps 10^{41} J per supernova,* and also that the frequency of eruptions of supernovae is about once in 30 years (say 10^9 s). This indicates a rate of injection of energy into the electron component within supernova remnants in the galaxy as about 10^{32} W.

Too much must not be read into the fact that this figure is precisely that calculated for the steady-state energy requirements of the galactic electron component. Much of the energy of energetic electrons within supernovae must be radiated still inside the source-volume. But over and above this, such calculations are very much of orders of magnitude rather than of precise quantities. Ginzburg himself once emphasized the sort of accuracy to be read into any calculations of

* For the Crab it is certainly larger, at least 10^{42} J.

this kind by giving the most important relevant equation, $1 = 10$, pride of place!

The electron situation regarding supernovae has a fairly sound basis: when we turn to the positive (nuclear) particles the difficulty is that there is no direct evidence of these, accelerated, within supernova systems. However, present estimates of the total energy output in supernovae are of the order 10^{43}–10^{44} J, and the corresponding average rate of energy release in the galaxy is 10^{34}–10^{35} W. Thus while there is no positive evidence that supernovae are the main source of the (positive) cosmic ray primaries, this is certainly not ruled out by the present estimates of the total energy balance.

It is interesting to consider what contribution might be expected to the present day cosmic-ray flux at the earth from the historic supernovae, which have been striking objects in their time; the Crab, seen to erupt in 1054, and the two famous supernovae of the 16th and 17th centuries, Tycho's star in Cassiopeia in 1572 and Kepler's star in 1604.

The Crab is estimated to be about 4×10^{19} m from the earth, so light takes about 4000 years to reach us from it. Light first arrived 900 years ago, and therefore charged particles from it can only be reaching us now if their diffused path is not greater than the direct path of light by more than about 25%. It is most unlikely that any of the particles which reach the earth fully diffused within the galaxy can meet this condition, and so it appears that the main cosmic ray intensity cannot yet be affected by the Crab eruption. The same exclusion, but even more decisively, applies to the two later notable supernovae.

6.4. *Pulsars*

The Crab, being near and of a convenient age, is the supernova remnant about which we know a great deal more than we do about any other, rather as we know far more about our own sun that we do about any other of the 10^{10} or so closely similar stars in the galaxy. What is not clear is whether the Crab is a typical supernova, for some of the features measured can be compared with those of other events, and in these there is enough evidence to suggest that the Crab is in some ways unusual.

When pulsars* were discovered several years ago, their evident properties of being small and rapidly rotating immediately brought

---

* Pulsars were first reported in 1968, and were so named because they were observed as sources of radiation coming in short sharp bursts with an extremely rapid repetition frequency. They were first observed at radio-frequencies but examples have now been detected in optical and X-ray radiation. They are considered to be 'neutron stars', condensed until the whole mass is of more or less the density of nuclear matter.

102

them forward as possible cosmic ray sources. A great deal of attention was therefore given to the properties attached to possible small massive sources, particularly in their interaction with surrounding plasma. Since these must all have originated in some sort of collapse, any significant angular momentum associated with the pre-collapse object must lead to extremely rapid rotation of the product as this becomes very small, and this is just what is observed for pulsars.

In 1970 a pulsar was found to be located within the Crab nebula, and this became one of the most closely studied of pulsars: its characteristic pulsation form was observed at radio, optical and X-ray frequencies, and its relationship with the Crab remnant as a whole as it is at present observed became evident. It is not yet at all clear that any pulsars can be thought of as independent possible sources, and in particular that theories of emission from them can be realistically developed which do not involve the special environment which is so obvious in the Crab and which must probably have arisen for all or most of them at the time of their collapse.

6.5. *Interpretation of the charge spectrum*

We have already (p. 61) given a brief description of the charge spectrum of cosmic ray primary nuclei as they reach the earth, and indicated how this carried information both about the composition of cosmic rays at their ' source ' and about the fragmentation which must have taken place in transit to us, and so about the conditions in the interstellar gas through which they have come. Increasingly detailed studies of this material have been a feature of work in the last few years, and some interesting conclusions have followed.

At a very early stage it was recognized that light elements like lithium, beryllium and boron, which are not expected to be present in detectable amounts in regions where nuclear activity is important were present in cosmic rays in quite substantial amounts (fig. 23) and it was supposed that these had come into being during transit and as a result of collisions of other (heavier) nuclei with the interstellar gas. A rough estimate of the thickness of this material which would have to be penetrated to lead to the formation of light nuclei to this extent was 4 g cm^{-2}, and the elaboration which has been made under more detailed scrutiny does not contradict this first simple estimate. In discussing the situation we are in the difficulty of having already taken the term ' primary ' as distinguished from ' secondary ' to describe what is incident upon the earth: now, to avoid misunderstanding we describe what leaves the source as ' primordial ', and what reaches the earth as ' surviving primordial ' and ' fragmentation products '. Thus ' primary ' is the sum of ' surviving primordial ' plus ' fragmentation products '.

The amount of fragmentation which appears to have taken place corresponds to the thickness, $4\,\mathrm{g\,cm^{-2}}$, of interstellar gas already referred to, and for a near-relativistic particle this corresponds to a lifetime of a few million years. This lifetime, however, depends on the assumption that the particles spend all their existence in interstellar gas of the density which is normally assumed, and it can be argued that if they had divided their time between movement in this medium and movement in very much less dense gas in the halo-like region, the actual lifetime in years might be much longer, perhaps as much as 10^8 years. It would be a most important step forward if it were possible to make a direct measure of lifetime as well as the indirect value deduced from the amount of matter the particle has moved through. One such measurement has been recognized for some time, in the sense that a promising method has been identified, but the results are not yet in the state of precision or agreement between different working groups which allows any firm conclusion.

The isotope ^{10}Be is unstable, with a lifetime at rest $\sim 4 \times 10^6$ years, decaying to yield ^{10}B, and the question is, does the ratio Be/B in fragmentation products suggest that ^{10}Be has or has not mostly decayed? The problem is manifestly complicated, apart from the actual measurements. Fragmentation nuclei of ^{10}Be will have differing Lorentz factors. Confident estimates of the actual production of Be and B isotopes by fragmentation are needed, and although these mostly arise from the break-up of the relatively abundant primordial carbon and oxygen, ^{10}Be in fact arises mostly from a second fragmentation of ^{11}B in grazing collisions with interstellar hydrogen. Better measurements of the Be/B ratio are to be expected, but the issue is likely to become still clearer when the actual proportions of the various isotopes of Be and B in the fragmentation material come to be measured. Meanwhile, if anything, the present measurements suggest a short lifetime, and are consistent with the containment up to detection of primary particles in the main interstellar medium of the disc.

The order of magnitude of the lifetime of cosmic ray particles from source to observation at the earth, several million years, is very large compared with direct traversal times within the galaxy, even that across a diameter. Accordingly the diffusion of these particles is not like that of particles spreading out indefinitely from a source through a diffusing medium of infinite, uniform extent. Rather, it is diffusion within a box of quite limited extent, and the characteristic ' age ' of the particles which fall as primary particles on the earth is determined by the extent to which the diffusing particles find their way out of the galaxy. This is the ' leaky box ' model of galactic containment.

In describing the mean thickness of interstellar gas which the cosmic ray particles have traversed as about 4 g cm^{-2} we do not, of course, for a moment propose that each individual particle has been travelling for approximately this same time. There must be a considerable spread of lifetimes, and in the absence of better evidence the 'leaky box' model suggests that the lifetime distribution should be considered as exponential, describing a situation in which each time a particle approaches the boundary of the box its chance of leaking out is the same, irrespective of past history. This description, of course, parallels that of the process of α-radioactivity. A form which appears to work well, given by M. Shapiro and his colleagues at the NRL, Washington for the probability $P(x)$ of a path length x is:

$$P(x) \sim x, \qquad\qquad x < 1$$
$$P(x) \sim \exp{(-0\cdot23x)}, \quad x > 1$$

where the mass thickness traversed, x, is expressed in g cm^{-2}. That there should be some lower limit to the exponential form is certain, since there comes a distance within which sources would certainly lead to anisotropies which are not found. The value $x = 1$ chosen by Shapiro is rather arbitrary, but generally seems reasonable. For the moment we imply that the path length distribution defined here does not vary with energy or energy per nucleon.

To describe the primordial composition of the cosmic radiation from which the primary particles which reach the earth has been derived, we have to know in as much detail as possible what happens in fragmenting collisions. The intensity of any nuclear species reaching the earth will consist partly of primordial nuclei which have survived and partly of other nuclei which are the product of such fragmentation from all heavier nuclei. At the present time, detailed knowledge about fragmentation is growing very quickly, and so the most reliable form of the primordial composition at the time of writing is likely to be superseded as further fragmentation data become available. On the whole, however, the changes still to happen are not expected to be dramatic.

In fig. 36, the left hand side of the diagram describes the composition as the particles reach the earth, while the right hand side shows the corresponding primordial composition. One notices immediately that the most abundant primordial elements have lost intensity in transit, while others of very low abundance, or indeed of no abundance at all, in the primordial radiation have mainly been formed by fragmentation. The most obvious of these are the elements of the Li, Be, B group: it is not questioned that these, as they reach the earth, are entirely the result of fragmentation, and the actual abundance of the elements is the basis of the estimate of the effective mass thickness traversed by the particles from their source.

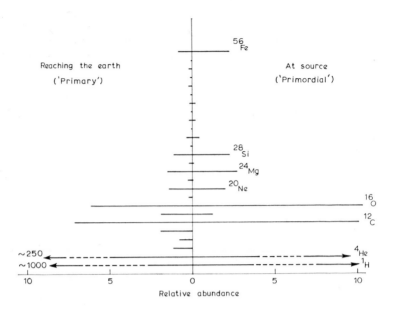

Fig. 36. Relative abundance of elements up to iron (56) reaching the earth and, derived from this, at source. All the elements which are important at the source suffer a considerable amount of fragmentation; only about one third of primordial iron nuclei escape fragmentation. At the other extreme, elements of very low abundance in the primordial radiation, or even absent in it, are produced in acts of fragmentation and reach the earth in considerable intensity. The abundances at the earth of hydrogen (protons) and helium (α-particles) are indicated on the scale of the remainder of the diagram: it is not possible to give different, meaningful, figures at source.

The primordial composition is very characteristic: apart from hydrogen almost all the nuclei of importance are of the very stable form with an even number of both protons and neutrons. This is entirely consistent with the view that this is matter ejected from nuclear synthesis in a collapsed star and its subsequent explosion, and the actual composition among these stable types may be expected to give more detail about these events. So does the fact that a few nuclei which do not conform to the even–even pattern—nitrogen, sodium, aluminium, probably phosphorus but almost certainly not fluorine occur, although only to a slight intensity. A variety of detailed theories are current about such sources: what is largely agreed is that stars with this sort of sensational history are massive, of mass at least several times that of the sun. It is a matter of speculation how this particular mix of elements, formed in a dense environment can get out with little or no ' source fragmentation '.

106

Up to now we have followed the assumption that the charge composition of cosmic ray primaries does not change with energy: this assumption seemed plausible from early observations (fig. 2) and in this section we have implicitly adopted a path length expression which is also the same for all energies. Detailed work over the last few years, and specially the extent to which reasonably accurate data have been extended to energies above 100 GeV/nucleon has, however, provided evidence that these assumptions are not justified over a wide range of energies. There seem to be two variations:

(i) The heavier elements become important with increasing energy/nucleon, and,

(ii) the amount of fragmentation gets smaller at higher energy/ nucleon.

The first of these variations is illustrated in fig. 37 (*a*) where data from five different experiments are plotted. The various experiments are not separated in this diagram, but all cover a large part of the energy range. The second involving three experiments under similar conditions is shown in fig. 37 (*b*). Here Li, Be and B are, of course, all fragmentation nuclei while C and O are mainly primordial. Nitrogen might suitably have been missed out of the diagram, since only slightly more than half of what reaches the earth is identified as fragmentation material, but this adjustment would not in fact alter the general picture which is shown here.

A reduction of the amount of fragmentation material with increasing energy is not difficult to understand in terms of the 'leaky box' model. Diffusion, as seen from the earth, remains complete, but the actual path lengths between successive reflections at the walls become shorter: the number of opportunities to escape per unit path length becomes greater, and the average lifetime before escape (or before observation) diminishes; so does the amount of accompanying fragmentation material. There is no reason to consider any complication at sources to explain this feature. The increasing proportion of heavy (iron-like) nuclei relative to the medium-mass nuclei (C, N, O) may be a secondary feature arising from this energy-dependent path length postulate, but explanations of this effect, which has only recently been established, have not really been worked out.

Although the stage of completely convincing explanation has not yet been approached, enough has been set out in the last few pages to indicate that the detailed study of primary particles in the energy range $0 \cdot 1 < E < 100$ GeV per nucleon, covering, as it does, the main energy-carrying members of cosmic ray primaries, is an area in which progress is rapid: it is an area, too, where different groups of workers, using considerable variations of technique, are finding consistent data.

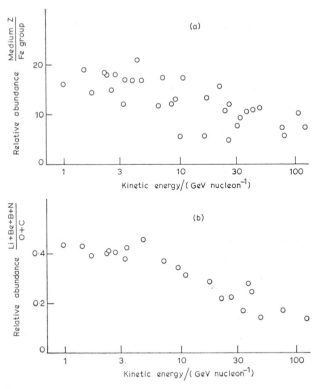

Fig. 37. Variations of the relative abundance of primary nuclei as a function of energy per nucleon. (a) Compares the ' medium Z ' group (mainly oxygen and carbon) with iron (from which its immediate neighbours are not distinguished). The energy scale is logarithmic and suggests that the proportion of iron becomes larger at energies of the order 100 GeV/nucleon. (b) Compares light elements, considered to be almost entirely the products of fragmentation (but see text for comment on the inclusion of nitrogen), with carbon and oxygen which are probably almost entirely primordial. The evidence for a reduction in the proportion of fragmentation products at higher energies seems well-established.

6.6. *Isotropy*

To describe an experiment as showing the direction of arrival of cosmic ray primaries at the earth as ' isotropic ' (coming equally in all directions) is usually to claim something considerably more general than has in fact been observed, for as a rule comparable measurements have not been made in all directions. What has normally been done is to observe over a band of directions defined as the apparatus scans because of the rotation of the earth, and since the solid angle ' seen ' by ground-level apparatus is always rather limited this band effectively covers only part of the whole celestial sphere. If observations have been made with comparable apparatus at stations situated over an overlapping range of latitudes, and all, within their respective zones of coverage, detect no variation of intensity, then one can very nearly consider that strict isotropy has been established. Even here, however, there is a weak spot, for along the rotation axis of the earth a scanning system is always looking at the same portion of the celestial sphere. Rather loosely, ' isotropy ' has come to mean an absence of any detected anisotropy (difference of intensity in different directions).

In this chapter we are concerned with isotropy relative to our galactic surroundings, and so scanned in sidereal time. For solar problems, isotropy is used in a very restricted context. Solar modulation is about the solar environment, and the earth (and up to now space probes dispatched from it) are constrained to a region near to the solar equatorial plane. A great deal might be learned from experiments carried out near the spin-axis of the sun, but we cannot yet get apparatus there.

If particles are said to form an isotropic distribution, this property of isotropy must be relative to something, and if there are no sources near enough and strong enough to the earth to produce a detectable anisotropy, we would expect the primary particles to be diffusing isotropically relative to the interstellar medium near to the solar system. This conclusion is the culmination of an historic approach to the general concept of isotropy.

6.7. The Compton–Getting effect

Ever since it was recognized that cosmic ray primaries have energies much larger than particles from terrestrial sources, people have accepted that their pattern of motion is certainly not imposed by the earth (although it is distorted, in a way now well understood, by the geomagnetic field). If the particles are diffusing isotropically it is certainly not with reference to the earth as a standard of rest.

This problem was first stated specifically by A. H. Compton and I. H. Getting in about 1935. Their examination of the problem pre-dates any understanding of the important part which the medium which separates astronomical objects plays in astrophysics. The space between planets and that between stars (and of course that between galaxies) was considered ' empty ' and quite without any effect on particles (even charged particles) moving through it. They envisaged rectilinear propagation and, given this, postulated that the frame of reference in which cosmic rays are isotropic cannot be in significant rotation relative to the distant galaxies, a conclusion which follows if no mechanism exists to ' entrain ' such particles near local features. They compared such effects as the rotation of our own galaxy, the motion of the sun within it, and of the earth in orbit, and concluded that ' if the cosmic rays come uniformly from all parts of the remote cosmos, our speed relative to their source is probably about that of the galactic rotation '.   From this postulate, a sidereal diurnal effect (with maximum when the observer on earth is looking forward along the direction of galactic rotation) of about 1.2% (crude) follows. But when this is modified for geomagnetic effects in middle latitudes, for the transition through the atmosphere to the secondary radiation and for the characteristics of the detectors then in use, it is reduced to an effect of only about 0.1%, with maximum near 20 h sidereal time.

   The work of Compton and Getting is important because of the way in which it illustrates the transformation of our thoughts about ' space ' in a time of only 40 years, but not less so because it provides what is essentially the correct approach to the problem relating to an isotropic cosmic radiation whatever the frame of reference within which it is isotropic may prove to be.   It is an interesting coincidence that the velocity of the earth relative to the remote galaxies was estimated in Compton and Getting's paper as 300 km s^{-1}, a velocity which is almost the same as that of the earth relative to the solar wind (which is something which can now be measured, but which of course refers to a reference frame co-rotating with the sun and leading to a solar rather than a sidereal effect).

   The non-solar Compton–Getting effect for which measurements can be attempted is accordingly that of the earth with cosmic radiation isotropic to the interstellar plasma in the neighbourhood of the solar system, and plausible assumptions suggest that this might be of amplitude about one tenth of that for the effect originally predicted in the treatment given by Compton and Getting.   This amplitude, perhaps 0.01% for ground detectors in middle latitudes, is very small indeed, but measuring it is made still more difficult (*a*) because a spurious sidereal effect may be generated from the relatively large solar diurnal effect as scanning directions on the earth look in varying directions relative to the equatorial plane of the sun with a modulating period of one year,* (*b*) because a similar spurious effect will arise if the solar effect is incorrectly processed for atmospheric conditions such as pressure and temperature, as these include modulating seasonal variations of effective period one year and (*c*) because the fine structure of solar wind modulation of cosmic intensity (see, for example, page 49) introduces a relatively strong ' noise ' signal against which the extremely small sidereal amplitude is to be detected.

   The possibility of these measurements has attracted a great deal of work, but the results emphasize the difficulties of measuring so small an effect.   There is general agreement that the most promising

---

* The difference frequency between solar and sidereal time is one cycle per year.

conditions are to work underground with a rock mass thickness of between 5 and 10 kg cm^{-2}. Fewer particles are present, so the statistical problem of counting enough of them is much more difficult, but they originate from much higher energy primaries, so the reduction from the crude Compton–Getting sidereal effect to the very discouraging surface effect is significantly mitigated by the reduction of the disturbing local features. The differences between results from various experimenters are puzzling and prevent any clear conclusion to be drawn at the moment. We may deduce that there is no large anisotropy near the solar system: what is not known is exactly what condition of the interstellar gas controls any anisotropy which may emerge from these various experiments.

6.8. *Isotropy at higher energies*

Underground measurements discussed in the last paragraph refer to a fairly narrow band of energies centred on about 10^{11} eV: this seems about the conditions under which any effect very close to isotropy would be most effectively studied: at lower energies spurious effects are too serious, at greater energies the number of primaries counted tends to become smaller. For example at ten times the depth, what was observed would derive from primaries near 10^{12} eV, and these would be even less liable to spurious effects, but the intensity would be much more than 100 times lower and working conditions would be long and extremely trying. Equivalent information is not feasible.

Much beyond this energy other techniques have to be used and, in brief, no certain departure from isotropy has been identified. Some ' feel ' for the data can be gained from what is known about the very highest energy primaries, for which the likelihood of non-galactic origin has to be taken seriously.

Isotropy at about 10^{17} eV has been studied at the Haverah Park site (p. 73), where data in a single band scanning a fixed part of the celestial sphere have been collected over a period of years. For this band about 10^5 showers from primaries near 10^{17} eV have been included, and isotropy is established to within about 1%.

As this chapter was written J. Linsley and A. A. Watson have published a survey of showers from primaries of energy estimated by them as all deriving from primaries of energy greater than ' a few times 10^{19} eV '. They have assembled material from three major shower arrays (one operated by the University of Sydney in Australia, one, relatively near the equator at Volcano Ranch, New Mexico, and a northern array at Haverah Park) and these in conjunction provide near-spherical coverage. At the limited accuracy of what emerges (fewer than 100 primary particles contribute to the whole distribution) inter-calibration of arrays is not a serious problem, and the result is shown as an 'equal area' plot in fig. 38. The U-shaped heavy line superposed on the diagram marks the galactic equator.

The results summarized in this diagram will probably be accepted by the reader as quite consistent with complete isotropy but consistent also with a number of decidedly anisotropic distributions of arrival directions. All that can be stated with confidence is that it does not show a distribution which can be plausibly related to a very few dominant source directions within the galaxy from which these primaries come along almost straight paths. It was easy to understand that the isotropy of low energy primaries arises because they are heavily diffused in small volumes of the galaxy, and as far as we are concerned, in that around the solar system. At the highest energies,

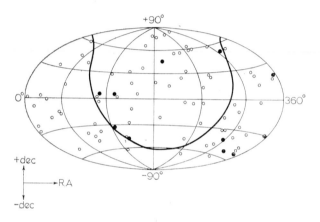

Fig. 38. Distribution, given in mid-1974, of 87 of the most energetic known primary particles (E_p > several times 10^{19} eV) by J. Linsley and A. A. Watson in celestial coordinates. The projection (of the whole sphere) used has the property that any area on the diagram is proportional to the solid angle in celestial space which it represents, and, to sufficient accuracy all parts have had similar exposure. For comments on the distribution, see the text. The bold full line marks the galactic plane.

the information in the Linsley–Watson diagram is not so easily related to diffusion without reference to the distribution of sources. If, as seems probable, these particles are not part of the cosmic rays contained within the galaxy, we start knowing hardly anything about the diffusion properties of extra-galactic space, but we do know that the extra-galactic universe involves a very uneven distribution of galaxies and that most of our near neighbours lie in more or less the same general direction, towards the great Virgo cluster of galaxies. As more data about very energetic primary particles accumulates, this is one direction to which particular notice will be given, to determine whether

112

there are any indications of preferential arrival from these important near-neighbours.

6.9. *The primary spectrum near* 10^{15} eV

The primary spectrum, as earlier sections have shown, can be represented on a double logarithmic plot over one range from perhaps 10^{10} eV to 10^{14} eV and another from 10^{16} eV to 10^{20} eV by two sectors of almost uniform slope, the second being of index $\sim 0 \cdot 6$ greater (i.e. steeper) than the first. If we accept this representation, there remains a most puzzling question about what is happening to the primary spectrum between the upper limit of the low energy range and the lower limit of the upper energy range, that is to say, at energies near 10^{15} eV. The question is particularly important because this change is the only obvious large feature in the whole known spectrum, and it is frustrating to have to record how little it is understood.

The real difficulty is observational, for it is about at this energy that the last few isolated direct records, mostly in emulsions, of individual primaries fall, while at the same time the properties of shower formation, which can be used to good effect at higher energies, are only partially helpful. The shower from such a primary will, at maximum development, contain fewer than a million secondaries, and sampling measurements either involve altogether too few particles or concentrate in the very central region, where shower growth is particularly irregular (fig. 29). Altogether, there does not seem to be any really good way of measuring energies in this region.

For the time being, nothing hard is known about this feature. It may mark a major change in the nuclear physics of the first collision in the atmosphere, it may arise from a special component of limited energy range superposed on a general spectrum which would itself be smoothly varying, it may show that the composition of primary particles changes sharply here, or that the galactic magnetic field is of such a form that it is around this energy that the general mode of motion in which the trajectories are guided within the galactic disc alters abruptly. All of these utterly different possibilities remain in court until something much more exact, and much more consistent from one group of experimenters to another, is achieved.

However, an energy of 10^{15} eV is a long way beyond the range within which the great bulk of cosmic ray energy is to be found, and the lack of understanding which surrounds the situation here can take its place with the uncertainties relating to the rare, very high energy particles, and probably without any bearing on the basic energy situation.

CHAPTER 7
cosmic rays: galactic or extra-galactic?

7.1. *The galaxy and its environment*

IN the previous chapter, we have implicitly written of cosmic rays as a phenomenon within the galaxy: we have worked out the energy balance inside it and described it as a ' leaky box ' from which particles can escape.   However, we have also hinted at the likelihood that the most energetic particles of all, which since they do not carry any significant part of the total energy cannot be argued about in terms of the energy balance, must be seriously thought of as coming in from outside the galaxy.   At the time of writing the whole broad picture of cosmic ray origin, galactic or extragalactic, is a major area of controversy, and it must be one task of any book about the subject, where so many uncertainties remain, to set out the current position on this problem which is quite fundamental in its contribution to cosmology as a whole.

At first sight, the view of cosmic rays confined within the galaxy has great attractions.   The rate at which energy has to be released is not unreasonable, and sources capable of yielding something not far short of this quantity have been identified, while the composition of matter from such sources about corresponds to what is found.   It is true that a detailed description of the acceleration process at such sources has not been achieved.   Further, the containment property of galactic magnetic fields and the density of the interstellar gas link quite convincingly with the lifetime of positive nuclei, as revealed by the fragmentation which has taken place, and with the radio-emission observed from the measured electron component.   Only at the highest energies yet studied does it seem that there may be a need for supposing that what we observe are particles from outside our galaxy.

The analogy with starlight, in the sense that practically all the starlight which reaches us is from stars in our own galaxy, is not a good one.   Optically even within the galaxy what we actually see depends very much on the obscuring properties of interstellar dust; in particular we cannot see anything at all of the main concentration of stars near the centre of the galaxy.   In contrast, we showed earlier (p. 92) that although particle trajectories may be very much deflected by interstellar fields, the absolute cut-off by dust does not hold for

114

atomic nuclei (or for electrons) in the same way as it does for visible light.

The reasons for thinking that the most energetic particles come from outside the galaxy are two-fold. The first is that the degree of isotropy already suggested by the data given in fig. 38 or, if you like, the form of what anisotropy it appears to support, is hardly what one would expect for particles from galactic sources. Even assuming the highest values of interstellar fields which can well be possible, the paths of these extreme energy particles are not going to be far different from straight lines. But the amount of galactic matter ' seen ' from the earth in the direction of the galactic pole (perpendicularly outwards from the plate-like form of the galaxy) is several orders of magnitude smaller than that ' seen ' in directions in the plane of the galaxy, and particularly than that towards the galactic centre. It would be extraordinary if this very large difference did not show up as anisotropies even in the limited data of fig. 38, the more so since there is not the slightest reason to suppose that the galactic matter in the polar direction contains an above-average concentration of sources: if anything, one would rather expect the opposite.

The second reason is more general. During the succession of startling discoveries which characterize the last twenty years of observational cosmology, it has become clear that our own galaxy is quiet and comfortably inactive as compared with many striking extra-galactic objects which are now being intensively studied. Whatever the detailed mechanisms which have accelerated cosmic ray primaries may be, it would be surprising if some of these regions were not capable of imparting far more energy to individual particles than any of the recognized source-objects in our quiet galaxy. Providing that they can get in, and providing that there are not reasons which could prevent them ever reaching us, the most energetic particles should surely come from outside.

The question of getting in is hardly a serious problem; the ' leaky box ' could probably leak as easily inwards as outwards. That as to whether they will ever reach us is more speculative. Particles from extragalactic sources must certainly have been travelling for a very long time, and the possible interaction (with loss of energy) of these particles of extreme energy with the cold black-body radiation, temperature about 3 K, which is predicted as arising as one of the products of the ' big bang ' model of the known universe (p. 88) is of critical interest. Up to now, there is no real evidence that even the most energetic primary particles have suffered loss or degradation in such interactions, but too much should not be read into this negative observation. Any major effect would be expected at energies near to the highest which are at present thought to be observed, and for these only a very few particles have yet been seen; also there is still

115

some flexibility in the calculations of possible interactions and the energy at which their onset ought to become significant. Further, the distances which these particles would have to travel for such effects to become unquestionable are a good deal longer than the distances to the nearer important extra-galactic features. The fact that no such effect has yet been recognized is accordingly not serious evidence that this interaction does not happen. On the other hand views about the structure of the galaxy would have to be strained to the limit to sustain the belief that even the particles of extreme energy are galactic in origin and contained in it.

The question which has now opened to serious discussion is whether a considerably larger fraction of cosmic ray primaries not susceptible to 3 K radiation degradation might not also be extragalactic, and come from some of the major energy-sources which have now been identified. What are at issue are particles down to a considerably lower energy and not ' everything ', for if the arguments for extra-galactic sources of the extreme high energy particles are strong, that for a galactic origin at low energies is all but certain. There are rare occasions when particles causing secondary effects at ground level have certainly come from our own sun, and lower energy primaries observable only in apparatus used under high balloon or satellite conditions contain a solar element much more frequently: we have referred (p. 79) to particles accelerated in the neighbourhood of Jupiter!

A discussion of this problem must also as a first step involve the question as to whether concentration on a simple distinction between galactic and extra-galactic cosmic ray primaries, and ignoring even the gross details included in the structure of either region, is justified.

Regarding structure within the galaxy, it is probably justifiable to ignore detail. At the time of writing, the first serious information about conditions near the galactic centre are coming from γ-ray observation. There seem to be concentric zones of distinctly differing properties, but the quantitative aspect of particle acceleration, and more so the diffusing of any such particles into the outer galaxy (i.e. the region containing the spiral arms) are open questions. The outer part of the galaxy certainly has an obvious structure shown by the spiral arms and the very irregular clustering of visible stars within the arms (fig. 34). This is easily seen for other galaxies in many respects like our own, and very irregular clustering of matter is also directly detected as regards interstellar gas in our own galaxy from the observations of radio astronomers on the 21 cm emission line from neutral atomic hydrogen. However, there is no reason to think that boundaries between these various volumes are of a kind which can be crossed only with difficulty, and to spend time upon them in the present discussion would only be a distracting complication.

In the larger structure of the universe beyond our galaxy the situation is different. Concentrations of stars into systems more or less like our galaxy do not occur at random, but are noticeably and often closely grouped, and very obvious ' clusters ' of galaxies can be picked out. When, on closer examination what has been identified as a cluster is accepted as being only a part of a still larger grouping, we are led to the idea of a ' super-cluster '.

This tendency towards strong ' clustering ' seems to extend to the largest scale that we have been able to recognize up to now. There is an overall tendency for galaxies to be moving away from each other: this is the basis of the notion of the ' expanding universe ' and, in the reverse sense, of the view that at a certain moment in the past all matter was concentrated in a small volume from which it exploded in the ' big bang '. This description, which quite certainly has not resulted in individual atoms moving away from each other uniformly, equally does not lead to galaxies being found uniformly spread and for the same reason. The effects of gravitation within the expanding system are very far from negligible, and groups of galaxies formed randomly may well have clustered as a result of simple gravitational attraction. If this is how clusters and super-clusters have come to exist, the only special feature of their relationship would be this gravitational attraction, and in particular they would not represent any sort of coherent containing-structure for cosmic ray primaries. All space would be one large reservoir from which extra-galactic cosmic rays entering our galaxy would have to come. There has never been much enthusiasm for this view applied to the main bulk of cosmic ray primaries, because of the extremely large fraction of all energy which would have to be assigned to the kinetic energy of cosmic ray particles. Indeed it was precisely this high energy needed if cosmic ray primaries were, as first thought, universal which led to the consideration of limited volumes, and particularly to single galaxies, as volumes within which alone this large energy density would be built up.

The recent identification from the large radio detector array at Westerbork in the Netherlands of extended sources of radio-emission which seem large even compared with our local cluster,* however, open up an alternative view: that entities on the scale of super-clusters can have a long-term integrated history of their own, which may have involved catastrophic events establishing a cosmic ray population. If this form of development has any relevance to super-clusters it would lead one to expect some degree of boundary property, and the possibility of an initial injection and then later the containment of cosmic-ray-like particles which occupy this volume

---

* For example: Diameter of galaxy, 10^{21} m (30 kpc).
Dimension of (local) cluster, $\sim 3 \times 10^{22}$ m (~ 1 Mpc).
Dimension of the radio-object 3C 236, $> 10^{23}$ m ($\sim 5 \cdot 7$ Mpc).

rather than space in general. Since super-clusters occupy at most about 1% of space, this possibly would result in a quite significant reduction in the total energy requirement, though this still remains high.

The main bulk of cosmic rays (energy in the range 1–1000 GeV) seem to have passed on average through perhaps 4 g cm^{-2} of matter since their initial acceleration and if we suppose that they were accelerated and then rather efficiently contained inside the galaxy, this is the thickness of interstellar matter through which on average these particles travel before they escape from the galaxy. However, if we go to the opposite extreme and suppose that these particles were formed in and now fill at least the volume of the local cluster of galaxies and that what we observe are such particles which have leaked *into* our galaxy, the picture is remarkably little changed. Because of totally different time scales, the efficiency of containment in the cluster is not likely to be important, and because the gas-density of matter in intergalactic space is now thought to be very much below that inside the galaxy (~ 3 protons m^{-3}) the main thickness of matter which these particles would encounter before reaching us would still be that traversed *after* they had penetrated into the galaxy. Once inside, they would have the same sort of history as would particles initially accelerated within the galaxy, and it is very difficult to see criteria which would distinguish between these extreme postulates. Because intergalactic gas densities, and presumably the magnetic fields associated with these, are so low, secondary effects such as the existence of some sort of emitted radiation are not to be expected at a detectable level outside the galaxy. Thus we do not seem likely to be able to establish directly either the presence of the absence of even a substantial cosmic ray flux in these regions.

There are three features which on the whole must be thought of as supporting galactic origin for the main bulk and one which is difficult, if only slightly so, on this hypothesis.

The favourable features are: the actual magnitude of the energy existing as kinetic energy of cosmic rays, the detailed spectrum of ultra-high-charge primaries (i.e. those in and beyond the stability zone of the 'magic numbers', $N = 126$, $Z = 82$), and the estimates of the lifetime of an electron component of the observed density. The difficulty arises because if the main bulk of cosmic rays are galactic, it may be thought surprising that the observed energy spectrum does not show some identifiable change at the galactic/extra-galactic cross-over.

7.2. *The energy-balance*

The kinetic energy of cosmic ray particles is certainly known, at least to an order of magnitude, in the outer parts of the galaxy, and is not in conflict with the output from the most likely sources of primordial particles (p. 101). More importantly, this energy is very similar to that estimated to be present as magnetic field energy.

Such an approximate equality can, of course, be pure chance: as a matter of fact both of these quantities are about the same as the energy of starlight near the earth, and this further near-equality is almost certainly only chance. However in a system of charged moving particles in magnetic fields there is a linkage which makes a connection between particle energy and magnetic-field energy much more significant. When a weak beam of low-energy electrons is fired through the field of a strong electromagnet, as was done in the classical experiments of J. J. Thomson and others, the magnetic field is thought of as rigid and unchanging, and it is the electrons which are channelled along orbits which are calculated accordingly. At the other extreme, if we try to remove a magnetic field existing in a large block of copper, it is the highly conducting property of the block, and the mass to which this property is attached, which is thought of as fixed, and it is the dissipation of the field energy which has to conform. Somewhere between these extremes is the situation, much more like that in our galaxy, where neither the charged particle flux nor the field is plainly dominant and where each is responsive to the properties of the other. The mathematical analysis of such ' magneto-hydrodynamic ' situations, when magnetic fields are strongly linked with moving ionizing gas, is highly complicated, and we shall not consider even simplified models here.

A ' leaky ' mode in which the result of a particle approaching the boundary of the galactic field is unpredictable, and has to be expressed in terms of probability, has been identified by some workers as one result of near-equality of particle kinetic energy and field energy leading to wave-like movements at the surface of the galactic field and to instabilities of the surface which allow a limited prospect of particle escape *providing the particle density outside is very much lower than that inside.* Notice that this sort of leak is valve-like and does not work both ways, and that it describes essentially a situation where particles originate and are contained in their own galaxy.

7.3. *Very heavy primaries*

We have deliberately postponed our discussion of very heavy primary particles, that is to say those of atomic number about 75 and above, to this point. In the earlier section about composition (p. 106) the main interest was the modification in nuclear collisions of a ' primordial ' radiation by fragmentation to yield the primary particles at the top of the atmosphere. The important feature was then the effective mass-thickness of interstellar (and conceivably inter-galactic) matter traversed.

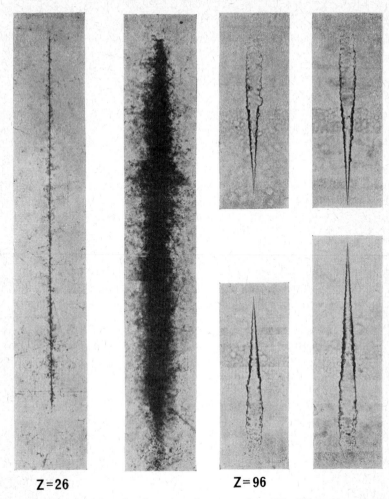

Z=26 **Z=96**

Fig. 39. Emulsion track of a super-heavy nucleus ($Z \sim 96$) compared with
that of an iron nucleus. This particular particle was also recorded in
plastic sheets in which the damage caused by its penetration can
subsequently be developed, and records from two sheets are shown.
The true length of the emulsion tracks is about 0·7 mm.

Very-high-charge primaries are rare, and so the total number
observed up to now is disappointingly small, but there is the advantage
that their tracks in sensitive material are almost impossible to miss.
The emulsion track on the left hand side of fig. 39 is that of an iron
nucleus. This is already some hundreds of times denser than the
track of a proton as has been shown in fig. 24. The track in the centre
is that of a very heavy primary, also in emulsion and on the same

scale. This would be easily visible to the naked eye, and it is difficult to imagine that it could be overlooked.

Iron nuclei provide the last well-established calibration point for emulsion tracks, so as one moves to higher atomic number there is plainly a critical problem of identification; it is only recently that the leading workers on this problem have become confident. We have here (fig. 39) the track of a particle of atomic number beyond ninety: ~ 96 is the best estimate. Over the last few years a new technique has evolved for recording these extremely heavy primary tracks and it has the very valuable advantage that the material is completely unaffected by the effects of light particles. Only heavy primaries make any sort of record, and this is therefore not cluttered with the effects from far larger numbers of light particles. The calibration problems of this new method are now being overcome, and it has become extremely important both used on its own or, as shown in this instance (fig. 39) in conjunction with an emulsion stack. In this method the structural damage caused by the passage of the particle through certain materials can be ' developed ' by etching agents to leave a conical tunnel into the material of the target, the dimensions of which under standardized conditions can be interpreted in terms of the velocity and the charge of the incident primary. Certain plastics in sheet form are suitable targets and in our example two sheets of this material, as well as the emulsion layer, were traversed by the primary. Each sheet here shows two etching ' pits ', nose-to-nose, since the etching process penetrates in development from both sides of the plastic sheet. The difference of ' pit ' produced in the two sheets illustrated corresponds to a slowing down of the primary between them: in this example the estimate is that when it left the second plastic layer, the primary had a residual range of about 1 g cm^{-2} and there was indeed more than this thickness of absorber before the next plastic layer of the pile, in which nothing was seen.

For our present argument the important members of the very heavy primary group are those corresponding to the ' lead group ' of elements, the heaviest group which shows long-term stability, and the ' uranium group ' all of which are unstable with various lifetimes. The lead group will have measured atomic numbers in the range about 80–85, the uranium group in that at 90 +.

The relative members of particles to be expected in these groups is well enough understood for them to be used as a ' clock ' which measures actual time (strictly the ' proper time ' measured by a device travelling with the particle in question) since formation. This is a quite different matter from the ' derived ' times which have been deduced from mass-thickness traversed, and is of the same nature as the ' ^{10}Be clock ' (p. 104). In spite of the rather few examples yet available, all the evidence from this particular approach suggests that

121

the lifetime of these particles, wherever it has been spent, is relatively short, only something like a million years, a figure which, if it is correct, is decisive evidence that these particular particles must be of galactic origin: they cannot have reached us from plausible extra-galactic sources. At the same time these particles have an abundance which relates coherently with that of lighter primary nuclei, those of iron and of others, and so if the very heavy groups are of galactic origin it is hard to suppose that iron and indeed most other primaries are not also galactic.

7.4. The electron component

The third feature that favours a galactic origin for the main bulk of primaries depends upon the extremely well measured data now becoming available about the electron component. The ratio of positive electrons to negative electrons has been one of the more difficult of satellite-mounted experiments yet accomplished, but the ratio is now well enough known to make it certain that some of the electrons are true primaries, although equally a considerable number are derived from the galactic population of nucleonic primaries as the ultimate product of various collision processes. It is generally agreed that the primordial electrons cannot be very old, since their loss of energy both in the 3 K radiation and even in collisions with starlight is so large inside or outside the galaxy that they would not survive. It has already been pointed out that the operation of these modes of energy loss would first show up as a strong steepening of the energy spectrum of these electrons, and although many observations have now been reported, there is no indication of such an effect.

There is accordingly very substantial, perhaps even overwhelming, evidence that the great bulk of primordial cosmic radiation had its origin within the galaxy, and what can be said against this evidence is weak. If most of the particles are of galactic origin and, as almost equally strong evidence suggests, those of extreme energy come from outside we would expect to see something recognizable at the transition from galactic to extragalactic radiation as particle energy increases through the primary energy spectrum.

7.5. The transition problem

The search for a transition has a long and fluctuating history, and the problem is made no easier because there is no known reason to expect it, within wide limits, at any particular energy. It must be no higher than about 10^{19} eV (total energy) and it can hardly be below about 10^{12} eV/nucleon.

In the early 1960s, the first observations of really large showers, from primaries of energy stretching well beyond 10^{19} eV, were carried out by John Linsley at Volcano Ranch in New Mexico, and these

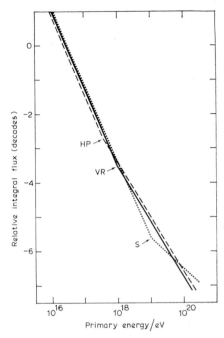

Fig. 40. The 'flattening' of the primary spectrum at extreme energies—a
cautionary tale! The full curve summarizes Volcano Ranch data
reported in 1963, with a significant change of spectral index near the
point marked VR. The broken curve shows Haverah Park data (1969)
with the change of spectral index near the point HP, and the dotted
curve, also of 1969 gives data from Sydney with the change of index at
the point S. The three curves, all involving different modes of measurement have been approximately normalized. *The change of spectral index
in all three sets is now believed to be spurious.* The form of the spectrum
now accepted is that given in fig. 26. While it is not possible entirely
to rule out some change near to 10^{20} eV, since there are so few showers
in this region, there is no reason to abandon the continuation of the
spectrum with little or no change of index to the highest energies yet
encountered (fig. 26).

showed what seemed to be exactly such a transition as might be
expected (fig. 40), with a 'tail' of lower slope (presumably extra-
galactic) emerging from the steeply falling spectrum (presumably
galactic) at high energies. Within a few years the other major groups
of workers with large shower arrays reported similar observations.
The worrying feature, however, was that while these observations
were qualitatively similar they were not quantitatively consistent, the
transition to a lower slope did not take place at the same energy in
the various experiments! Second thoughts and better analytical
techniques soon established that such an effect is exactly the way in

which defects of recording and of treatment of the data would manifest themselves. Now it has become evident that a great deal, and perhaps all, of effects like that shown in fig. 40 are not real properties of the primary but rather examples of data not well understood and imperfectly analysed.

The present situation is that very nearly all of the flattening tendency at the most extreme energies has, with much refined analytical procedures, been removed. It is an open question whether what remains is real or whether this also is artificial, arising from limitations of the whole measuring operation which have not yet been recognized. But however this question is answered, it does not seem likely that this region, and the few additional detected primaries there which may turn out to indicate a flattening of the spectrum, represents the transition from a low-energy galactic to a high-energy extragalactic flux. There are not enough of these particles.

It is very tempting to place the transition in a region of the spectrum which is difficult to study and where, therefore, specific evidence against any particular proposal is hard to establish. Such a region is that near primary energy 10^{15} eV, where the bulk of the evidence is that there is at least a rather accelerated steepening of the spectrum, but where for the time being experiments on the detail of what is happening are not consistent. That this should be the energy at which the galactic to extragalactic transition takes place has no stronger basis than ignorance, and two or three rival interpretations of the spectral features near this energy are generally regarded as more probable.

Perhaps the final point to be made about our failure to identify where the galactic spectrum becomes replaced by an extragalactic spectrum is one of caution. A 'simple' form of function such as that illustrated in fig. 26, where the spectral index is shown remaining unchanged for a range of more than a thousand in energy and more than 10^6 in intensity, invokes a 'simple' view as to what is happening, and the idea that 'nothing unusual' breaks the smooth progression over this enormous range of intensities. The use of log–log plots, which are essential if this wide range is to be accommodated in an understandable diagram leads to precisely this sort of over-simplification of thought. There can be no doubt that such a graph does bring out a most important trend, extending over energies differing by many powers of ten, but equally it could obscure features which are noticeable only in very detailed study within a relatively short range of energy. The graph in fig. 26 is not the result of a single coherent experiment; rather it is the best attained amalgam of varied experiments almost all different in detail and probably differing in reliability. We cannot exclude the possibility that somewhere in this featureless line as we now give it, there are changes which, if one elected to study some particular region in detail, would not only be detectable but perhaps also indicative of the transition that has proved so elusive.

7.6. Postscript

It is difficult to come to the end of an account such as that we have tried to give in this book without leaving the subject at a stage which emphasizes what is *not* known; in other words what is immediately posing problems of importance not yet solved. As a counter to this, we would remind the reader how very much has been found out. We know that most cosmic ray primaries are formed and are contained in the galaxy: we know their nuclear composition at their sources and we know the power output required within the galactic disc of such sources: we think we can recognize at least one kind of source. We also know a great deal about their propagation through interstellar space, and in some detail within the solar system. Although some critical features are still unknown, we have developed an extremely detailed understanding of the growth and ultimate absorption of the very complex secondary component which is developed from these primary particles in the terrestrial atmosphere.

In so doing we have found many points of contact with other approaches to cosmology, and often in ways where one approach confirms conclusions arrived at by others. One can hardly doubt, as other features come to be understood, that this interaction of modes of study will continue, and increasingly strengthen the overall structure of data and the ideas to which these lead.

APPENDIX A
magnetic rigidity

THE radius ρ of the circular trajectory of a charged particle moving in a plane perpendicular to a uniform magnetic field of flux density B is given by:

$$\frac{mv^2}{\rho} = ZeBv,$$

where m and Ze are the mass and charge of the particle, and its velocity v is small compared with the velocity of light, c. Thus

$$B\rho = \frac{mv}{Ze} = \frac{p}{Ze},$$

where p is the momentum of the particle.

Under relativistic conditions the expression $B\rho = p/Ze$ remains true, although p cannot still be replaced by mv: the ratio momentum to charge is then the 'magnetic rigidity'.

Since the (relativistic) total energy E is given by:

$$E^2 = (mc^2)^2 + (pc)^2,$$

when $pc \gg mc^2$

$$cB\rho = \frac{E}{e}\frac{1}{Z}.$$

This equation is correct in any consistent set of units, and in particular if the energy in electron volts is E_{ev},

$$E_{\text{ev}} = \frac{E}{e},$$

so

$$cB\rho = \frac{E_{\text{ev}}}{Z}.$$

If B is measured in tesla and ρ in metres, and since $c = 3 \times 10^8$ m s^{-1},

$$3 \times 10^8 B\rho = E_{\text{ev}}/Z.$$

(This expression will frequently be encountered with B in gauss, ρ in cm:

$$300\ B\rho = E_{\text{ev}}/Z.)$$

If we consider a particle of mass number A and energy E_{ev} per nucleon, then:

$$3 \times 10^8 B\rho = \frac{A}{Z} E_{\mathrm{ev}}.$$

For the proton $\frac{A}{Z} = 1$, for most light nuclei (C, O, etc.) $\frac{A}{Z} = 2$ and heavy elements $\frac{A}{Z} \sim 2\cdot5$.

Table. Radii, R, of the trajectories of particles of various energies, E, in various magnetic flux densities, B.

B/tesla	R/m		
	$E = 3 \times 10^{12}$ eV	$E = 3 \times 10^{16}$ eV	$E = 3 \times 10^{20}$ eV
10^{-10}	10^{14}	10^{18}	10^{22}
	(4×10^{12})	(4×10^{16})	(4×10^{20})
10^{-9}	10^{13}	10^{17}	10^{21}
	(4×10^{11})	(4×10^{15})	(4×10^{19})
10^{-8}	10^{12}	10^{16}	10^{20}
	(4×10^{10})	(4×10^{14})	(4×10^{18})

For each pair of figures the upper is for protons (or electrons), the lower (in brackets) for particles $Z = 25$; all are in metres.

The radii quoted are to be related to such quantities as:

Diameter of galaxy	$\sim 10^{21}$ m
Thickness of galactic disc	$\sim 10^{19}$ m
Distance earth–sun (the 'astronomical unit')	$\sim 1\cdot5 \times 10^{11}$ m
Interstellar field in galactic disc	$\sim 10^{-10} - 10^{-9}$ T.

APPENDIX B

the Van Allen radiation belt

Background

THE main magnetic field of the earth is closely that which would arise from a magnetic dipole located not very far from the centre of the earth. There seems to be no doubt that the source of this field is internal to the earth, but a completely convincing explanation of its origin has not yet been given. It varies slowly, the axis of the dipole seems always to be near the spin axis of the earth and the study of rock magnetism suggests strongly that the field has on many occasions reversed in direction, by passing rather quickly through a condition of near-zero field to one of the opposite polarity rather than by a field of more or less steady magnitude turning its axis through 180°.

Against this internal effect, sensitive field measurements outside the earth show a variety of small-scale changes which are often short and sharp. Nowadays many of these are caused by man, while a frequent cause is in the field changes induced by lightning strokes. An important and recognizable source of magnetic disturbances, however, is extra-terrestrial, and ' magnetic storms ' have been studied for many years. They have been related, in particular, with solar activity, and their relationship to active regions of the sun, and so with solar flares on the one hand and the eleven-year cycle of activity on the other, was well established before similar features came to be recognized in the cosmic ray flux. The fact that the main phase of a magnetic storm developed roughly one day after the solar flare with which it was correlated, and that the cosmic ray Forbush effect had an onset at substantially the same time made the close relationship between cosmic ray modulation and magnetic disturbances very clear.

A particular feature which came out of detailed study of the development and history of magnetic storms was the growth of evidence for the establishment during these events of a ' ring current ' in the equatorial plane and effectively distant several earth radii, but for a long time no mechanism for bringing this about was forthcoming.* However, in 1957 a young American, Fred Singer, drew attention to a

---

* The first evidence which seemed to call for the existence of a ring current appears to have been given by Carl Störmer as long ago as 1911–12, in his attempts to explain the position in latitude of the zones of maximum auroral activity.

category of 'trapped orbits' for charged particles in the geomagnetic field, which if the particle population was large enough, would exhibit the characteristics of a 'ring current'. The trapped orbits implied that particles in them would stay there indefinitely, and second order effects would have to be invoked to allow particles to get into this category of orbit in the first place and also eventually to get out.

A typical trapped orbit is shown in fig. B1, which illustrates the basic mechanism of trapping. A particle spirals along a geomagnetic field line where the form of the spiral is described by a 'pitch angle', α, the angle between the actual trajectory at a given point of the motion and the guiding field line. Following this particular field line, the geomagnetic dipole field is weakest in the equatorial plane and becomes stronger until the point of entry into the earth is reached.

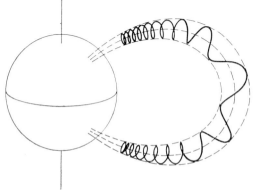

Fig. B1. Trapped orbit in the geomagnetic field: the pitch angle of the spiral has its minimum value, α_0, in the equatorial plane, increases away from the plane until it reaches $\alpha = \pi/2$ at the reflection points. The smaller α_0, the further towards the earth the motion continues before the condition $\alpha = \pi/2$ is reached.

Now the simple motion of the particle in this (static) magnetic field cannot involve any change in its energy, although the radius of gyration of its motion in the field must vary as the field increases or decreases. This situation requires that the pitch angle, α, varies with the field, B, and they are related by:

$$(\sin^2 \alpha)/B = \text{constant.}$$

Figure B1 shows that from the smallest value of B, say B_0, occurring in the geomagnetic equatorial plane, with pitch angle α_0, the motion is towards higher values of B with increasing pitch angle until the latter reaches $\pi/2$, where the maximum field, $B_{\max} = \dfrac{B_0}{\sin^2 \alpha_0}$.

Here, the orbit is instantaneously circular, but because the field is not constant, a residual force acts upon the particle in a direction perpendicular to this circular orbit, that is to say along the field line but now in the direction *towards* the equatorial plane. The particle has, accordingly, been reflected; it passes on through the equatorial plane, when its pitch angle will again be α_0 and is once more reflected in the opposite hemisphere at the point where α becomes $\pi/2$. The particle is ' trapped ' in an orbit which, in this simple statement, will return it backwards and forwards for ever. Trapped particles will in general have different values of α_0, and accordingly be reflected at different points along the guiding field line. If $\alpha_0 \to \pi/2$ the whole motion will remain in the immediate neighbourhood of the geomagnetic equatorial plane. The time to complete one cycle of this motion is a fraction of a second.

A final feature of the long-term form of such a trapped orbit is that because the guiding line of force is curved there is a lateral drift of the whole system around the earth which is in opposite senses for positive and negative particles (east to west for positive particles). It is this drift, which, were there enough particles in the trapped orbits, would give rise to the ' ring current '. The drift velocity is typically a few minutes per radian.

The important speculations about the trapped orbits were necessarily based on the view that there must be ways, perhaps only intermittently, in which particles could get in or out of them. Störmer's interest was to relate the regions, where containment might fail and particles pour out, with the auroral zones, which were his particular interest. Singer put forward the view that a large accession of particles into the ring current system might follow occasions of strong particle emission in solar flares.

Discovery

While the nature of trapped orbits was clear enough, it was for a long time uncertain to what extent there were really particles in these orbits, and indirect studies of this problem had not got far when the development of satellite vehicles allowed a quite direct approach by sending detecting counters right into the volume where such orbits must lie.

The first evidence confidently interpreted came from the satellites Explorer 1 and Explorer 3, but in hindsight it seems likely that moderate increases of counting rate on Sputnik 2 at altitudes of about 600 km arose in the fringe of the region which we are about to describe.

The Explorer satellites were on very eccentric orbits, which carried them from minimum altitude of less than 400 km to maximum beyond 2500 km. The counters installed upon them showed a definite but gentle increase to beyond 1000 km but then the increase was rapid and at about 2000 km the counters stopped working. The fact that this blank response was approached through an increase rate actually measured, and because it was not reasonable to suppose that this heralded a region in which there were no cosmic rays, pointed clearly to the interpretation that the radiation falling upon these counters was here so intense as to block their operation. Once discharged, the counters were never allowed to come back to their operating threshold and so counting stopped.

These measurements by James Van Allen, reported in 1958, led him to the view that there existed an intense radiation around the earth which could not approach nearer than about 600 km from the surface. This could only be a strong charged particle population controlled by the geomagnetic field, in fact in the 'trapped orbits' we have already described.

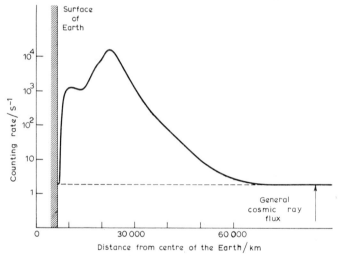

Fig. B2. Counting rate of a single Geiger counter carried on Pioneer 3 as it passed through the radiation belt (after Van Allen). The counter may be considered as omnidirectional in sensitivity, and the cosmic ray contribution to the counting rate is shown becoming substantially the total rate beyond about 70 000 km. It is of interest to compare this diagram with fig. 8: between about 70 km above the surface of the earth and about 1000 km the primary cosmic ray flux is again the main observable component, a feature which brings out very strikingly the limited range over which the atmosphere is important as compared with the region of strong influence of the geomagnetic field. Two radiation belts are indicated, but the actual rates involved are extremely complex arising from the varying spectra of electrons and protons in different parts of the belts and also the detailed sensitivity of the counter to the incident particles and to various secondary modes of discharge. The path of the satellite was *not* in the geomagnetic plane.

131

The early orbiting satellites were quickly followed by space-probes penetrating far from the earth, and within about a year it was established that the zone of intense particle radiation did not extend indefinitely, and that the outer boundary was somewhere about 60 000 km from the earth. Figure B2 shows the counting rate recorded for a Geiger counter in the probe Pioneer 3.

Other experiments established the nature and energy of the particles and allowed their distributions to be mapped. There seem to be two sections to the radiation belt (which is therefore sometimes described in the plural): these are perhaps to be identified in fig. B3, but it must be remembered that the precise form of the belt and its regions of high intensity depend very much on the details of detectors used. Protons and electrons are found in all parts, but the inner section contains particles of relatively high energy (about 100 MeV for protons) which the outer part is of much lower energy particles (for protons, a few MeV). The inner belt is also steady in form and intensity while the outer belt is much more variable. While details of origin are difficult to establish, it is probable that these contrasting inner and outer belts are from different sources.

The inner belt is probably largely the result of outward-moving neutrons, derived from cosmic ray interactions near the top of the atmosphere, decaying and so actually producing charged particles (protons and electrons) within the radiation belt. This form of injection is unique, since it does not involve any distortion of the belts or special mechanisms of entry.

In contrast, the outer belt, much more variable in its characteristics, probably depends upon the capture of solar particles, and it is when these are abundant that this belt and the ring current are at their most prominent. The injection of these particles in all likelihood does not offer very serious problems. While near the earth the geomagnetic field is closely that of a dipole, at distances of several earth radii it begins to become modified by boundary conditions between the volume controlled by it and that occupied by the solar wind as it presses upon the geomagnetic cavity and flows around it. Downwind from the earth conditions are confused, there is some turbulence and it is possible to imagine situations in which particles can be fed in and out of the radiation belts. It is this uncertain region which probably also determines the general outward limit of the radiation belts and it is probably here also that injection takes place. The relationship of ring-current and the aurora is also to be linked with this sort of situation.

The field in the region occupied by the inner belt is both much stronger than that several times further from the earth and also protected from external disturbances by the more flexible outer parts of the field. We have already indicated a mode of injection of particles into this region which does not depend at all upon field distortions and it would be satisfactory if there were a comparable approach to the question of the removal of particles from the inner band. This band would then carry a particle population determined by an equality of injection and removal. Such a mechanism can be thought of in terms of collisions in the extreme outer part of the atmosphere.

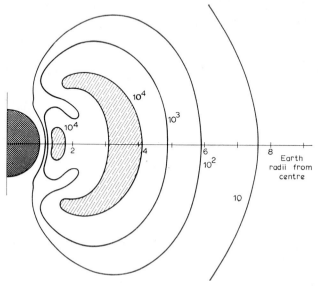

Fig. B3.   Section of the radiation belt on a scale of earth radii, as measured by a single omnidirectional Geiger counter. This is derived from counters on several vehicles, and necessarily involves a degree of simplification. The limitations as to *what is counted* indicated under fig. B2 apply also here.

Every trapped orbit of the kind illustrated in fig. B1 is nearest to the earth at the ends of the motion, when $\alpha = \pi/2$, and it is here also that the particle spends most time. So any interaction between particles and the atmosphere need only to be considered at these extreme ends of each orbit. Because the possible collisions take place when $\alpha = \pi/2$, the effect must almost always be to decrease α_0 for the particle in question, and then still deeper penetration in the atmosphere takes place to regions where another collision is now more likely. Hence once *one* collision with the atmosphere has taken place, the next has become more probable, and soon more, and the residual life of that particular trapped particle is likely to be very limited. Such a mechanism probably determines the form of the inner surface of the whole radiation belt complex in detail. Referring to fig. B3,

133

it is noticeable that the really high intensity of the inner band is restricted to near-equatorial latitudes. This is in accord with the mechanism outlined in this paragraph. Those particles will survive longest on any particular field line which penetrate least far into the atmosphere, and those which best meet this prescription are those with large α_0, which are constrained for their whole motion to remain near the equatorial plane.

Significance of radiation belts

The reader will have become aware that the problems posed by the radiation belt of the earth are, although complex in detail, not of fundamental importance. Recognition of the belt, and understanding of its operation have however assumed a much wider significance. Space probes on missions taking them close to the other planets will normally be equipped to detect particles of the kind encountered in the terrestrial radiation belt. The presence or absence of such a belt for other planets is most important evidence as to whether or not the planet has a significant (dipole) field.

The outstanding positive evidence comes from the Pioneer observations near Jupiter, for there is no doubt that around Jupiter there are very intense radiation belts of the Van Allen type. Although these have been under scrutiny for only a few hours, it is apparent that the situation is much more intense and violent than the corresponding phenomena about the earth. The internal field of the planet is much more complex than a simple dipole, and the whole system much less stable. It seems certain that the electrons observed near the earth which have recently been related to sources near Jupiter (p. 79) are accelerated in the outer regions of its belt structure.

INDEX

135

THE WYKEHAM SCIENCE SERIES

THE WYKEHAM TECHNOLOGY SERIES

All orders and requests for inspection copies should be sent to the appropriate agents. A list of agents and their territories is given on the verso of the title page of this book.

† (Paper and Cloth Editions available.)